Energy for Development

ENVIRONMENT & POLICY

VOLUME 54

For further volumes:
http://www.springer.com/series/5921

Ferenc L. Toth
Editor

Energy for Development

Resources, Technologies, Environment

Springer

Editor
Dr. Ferenc L. Toth
International Atomic Energy Agency (IAEA)
Department of Nuclear Energy
P.O. Box 100
Wagramerstrasse 5
A-1400 Vienna
Austria

DISCLAIMER
The views expressed in this publication do not necessarily reflect those of the authors' organizations, the IAEA or its Member States. The use of particular designations of countries or territories does not imply any judgement by the publisher, the IAEA, as to the legal status of such countries or territories, of their authorities and institutions or of the delimitation of their boundaries.
The mention of names of specific companies or products (whether or not indicated as registered) does not imply any intention to infringe proprietary rights, nor should it be construed as an endorsement or recommendation on the part of the IAEA.
The authors are responsible for having obtained the necessary permission for the IAEA to reproduce, translate or use material from sources already protected by copyrights.

ISSN 1383-5130
ISBN 978-94-007-4161-4 ISBN 978-94-007-4162-1 (eBook)
DOI 10.1007/978-94-007-4162-1
Springer Dordrecht Heidelberg New York London

Library of Congress Control Number: 2012937474

Printed on acid-free paper

Springer is part of Springer Science+Business Media (www.springer.com)

Preface

Energy has become the central theme in several fields of research and in various policy arenas, including international negotiations and agreements. Research has demonstrated that providing access to affordable energy is a key component of global and national efforts to alleviate energy poverty, improve social equity and promote development in general. The economic and physical sciences also indicate that the current global energy supply system raises many concerns, ranging from energy security and technological development to environmental impacts. In the international policy arena it is widely recognised that global coordination and cooperation is needed to support economic development through sustainable energy strategies and to avoid environmental calamities such as climate change. At the national level, many countries are in the process of revising their national development and energy strategies and redesigning the related government policies.

This volume brings together bold visions, in-depth analyses and innovative ideas about how to resolve the 'energy for development' predicament: how to meet the rising demand for energy for economic growth while reducing greenhouse gas emissions to avoid climate change. The chapters are based on contributions by a diverse group of leading scientists and prominent national and international policy leaders to an IAEA Scientific Forum devoted to the theme 'Energy for Development'. The material reveals the multiplicity of energy–development–environment problems and the diverse approaches being adopted to resolve them.

The first set of chapters provides an overview of the broad range of issues involved in the energy–development linkages, the main characteristics of past developments that have led to the current problems and the ongoing international efforts to find ways to resolve them. They look at various ways of rethinking the mechanisms through which development, energy and environmental aid is delivered from developed to developing countries, and the role of energy in achieving the Millennium Development Goals. Other chapters explore the increasingly important role of international institutions in addressing and solving the problems emanating from the development–energy–environment linkages; the importance of multilateralism for energy efficiency and clean production in global energy management; and the opportunities offered by existing international energy institutions in advancing energy for development.

The second cluster of chapters analyses the social, economic and technological drivers that will shape energy demand in the future. The main issues include demographic processes, such as rural–urban migration and the implications of the different lifestyles and household energy use on the demand for energy; the role of constructing and operating infrastructures and the ways short-term policies and decisions shape long-term energy demand; energy efficiency and its contribution to alleviating energy poverty and environmental impacts; and the importance of ensuring that environmental externalities in the energy system are reflected in energy prices and are considered in energy policy making.

The final group of chapters is devoted to energy and electricity supply. The chapters address constraints and opportunities for meeting the ever-increasing demand for energy by assessing the near- and long-term availability and costs of energy resources; government regulation and market organisation in the energy sector, especially in the electric power sector; the dynamics of energy supply systems and the importance of innovation and technological development; the increasing role of renewable energy sources and innovative ways of financing their deployment in the developing economies of South and East Asia; and the linkages between energy and food security, and in particular their implications for the poor. Additional chapters explore the prospects for two key baseload power generation technologies: fossil fuel plants equipped with carbon capture and storage, and nuclear energy. In 2011, many discussions of nuclear power plant safety were focused on the accident at TEPCO's Fukushima Daiichi nuclear power plant, caused by the extraordinary natural disasters – an earthquake and a tsunami – that struck Japan on 11 March 2011. The accident and subsequent safety reviews have had an impact on the prospects for nuclear power. Globally, the accident is expected to slow or delay growth in nuclear power but not to reverse it. Most countries with nuclear power have reaffirmed their commitment to expanding nuclear power while incorporating all the lessons to be learned from the accident. In countries considering the introduction of nuclear power interest also remains high.

The main objective of the book is to take the energy–development–environment discussion and policy formulation a step further. The book is intended to serve as resource material for scientists working on related topics in many disciplines, including economics, policy analysis, international studies and the environmental sciences. It might also be useful as background or reading material in postgraduate courses in these subjects. In the policy field, the target audience includes national energy planners and delegates to international negotiations on development (e.g. the UN Commission on Sustainable Development), climate change (the UN Framework Convention on Climate Change, UNFCCC) and other environmental agreements (Convention on Biological Diversity, Convention to Combat Desertification), as well as policy makers and their advisors. Decision makers in private companies would benefit from learning about recent trends in international development and environmental programmes to assess investment and other business opportunities.

Finally, experts and officers at international development, energy and environmental organisations such as the United Nations Development Programme (UNDP), the UN Department of Economic and Social Affairs (UN DESA), the United Nations Environment Programme (UNEP), the United Nations Industrial Development Organization (UNIDO) and others could use the material in preparation for the next Earth Summit, the UN Conference on Sustainable Development (Rio +20), to be held in Rio de Janeiro, Brazil, in June 2012, and in the final phase of the negotiations for the post-Kyoto agreement under the UNFCCC.

Hans-Holger Rogner
Section Head, Planning and Economics Studies Section
Department of Nuclear Energy
International Atomic Energy Agency (IAEA)
Vienna, Austria

Contents

List of Boxes

List of Figures

List of Tables

Energy for Development: A Key to Long-Term Sustainability

Ferenc L. Toth and Maria L. Videla

Abstract Understanding the broader context in which energy–development linkages are addressed is an important first step towards in-depth discussions of energy demand, supply and technologies. Energy-related aspects of international sustainable development targets, programmes and declarations provide increasingly important platforms to address these issues. Statistical causality analyses of the energy–development relationship explore why energy is important for social and economic development (measured as GDP per capita and Human Development Indices, HDI) but the current estimation techniques lead to inconsistent conclusions. The chapter also provides an overview of institutional, supply, demand, technology and environmental topics addressed in this volume.

1 Introduction

Energy has been a central concern to humankind throughout its history. The adequate provision of energy services has been important for economic development since the Industrial Revolution. In recent decades, energy issues have become a fundamental component of conceptual and strategic discussions on global sustainable development.

Energy is generally recognised as a key issue in achieving sustainable development. Several high-level conferences and declarations have confirmed that the provision of adequate energy services at affordable prices, in a secure and environmentally benign manner, and in conformity with social and economic development needs, is

F.L. Toth (✉) • M.L. Videla
Department of Nuclear Energy, International Atomic Energy Agency (IAEA), Wagramerstr. 5, P.O. Box 100, A-1400 Vienna, Austria
e-mail: F.L.Toth@iaea.org; M.Videla@iaea.org

F.L. Toth (ed.), *Energy for Development: Resources, Technologies, Environment*, Environment & Policy 54, DOI 10.1007/978-94-007-4162-1_1,

essential for sustainable development. Reliable energy services are the preconditions for investments that bring about economic development. They facilitate the learning that is crucial for developing human capital, and promote equity by enabling the less well-off to study and escape from poverty. Therefore, energy is vital for alleviating poverty, improving human welfare and raising living standards. However, the provision of energy services also raises other crucial sustainability concerns. The socially optimal depletion of non-renewable energy resources has been at the centre of the sustainability debate for decades. The environmental impacts of different energy forms and their repercussions for society – ranging from the damage imposed on socio-economic and material assets to risks to human health – can undermine the sustainability of development.

Many current patterns of energy supply and use are unsustainable. About a third of the world's population relies on the use of non-commercial fuels that have negative impacts on human health and the environment. Some 1.4 billion people have no access to electricity. Several regions of the world have no reliable and secure energy supplies, limiting economic development. The challenge is to design and implement energy development pathways that will support progress towards long-term sustainable development.

This chapter provides a background to the rest of this volume by presenting the broader context in which the linkages between energy and development are explored. Section 2 presents an overview of the energy-related aspects of international sustainable development programmes and declarations. Section 3 explores why energy is important for social and economic development (measured as GDP per capita and Human Development Indices, HDI) and assesses the recent attempts to conduct statistical causality analyses of energy–development relationships. Section 4 presents an overview of the issues addressed in the rest of the volume. Section 5 summarises the main points concerning energy poverty and development goals as well as policies for achieving them.

2 Energy and Sustainability

The report of the first major United Nations (UN) conference entirely devoted to environmental issues, the UN Conference on the Human Environment (UNCHE) in Stockholm in 1972 recognised that 'the protection and improvement of the human environment is a major issue that affects the well-being of peoples and economic development throughout the world' (UNCHE 1972, p. 1). The conference report, the Declaration of the United Nations Conference on the Human Environment, known as the Stockholm Declaration (UNCHE 1972), is dominated by concerns about the environmental implications of socio-economic development and the repercussions of environmental degradation for the development prospects of present and future generations. The energy sector is not explicitly mentioned in the Stockholm Declaration, but several items hold messages for energy production and use. Principle 5 of the document declares that 'non-renewable resources of the Earth must be employed in such a way as to guard against the danger of their future exhaustion and to ensure that benefits from such employment are shared by all mankind' (UNCHE 1972, p. 2). This presages the (still unresolved) sustainability dilemma of using non-renewable resources and the intergenerational equity element of sustainable development. Principle 6 speaks to the environmental implications: 'The discharge of … substances … in such quantities or concentrations as to exceed the capacity of the environment to render them harmless … must be halted in order to ensure that serious or irreversible damage is not inflicted upon ecosystems' (UNCHE 1972, p. 2). These are early indications of the critical load concept and the precautionary principle that have become key concerns in the environmental dimension of sustainable development.

Fifteen years later, the report on *Our Common Future* by the World Commission on Environment and Development (WCED 1987) describes sustainable development 'as a process of change in which the exploitation of resources, the direction of investment, the orientation of technological development, and institutional change are all in harmony and enhance both current and future potentials to meet human needs and aspirations.' The Commission's report describes the challenges involved in meeting these goals and recognises the importance of energy in sustainable development by devoting one of the six 'challenges' chapters to this issue. Starting from the premise that development crucially depends on the long-term availability of energy 'in increasing quantities from sources that are dependable, safe and environmentally sound' (WCED 1987, p. 168), the Commission defines four elements of sustainability for energy use: sufficient growth of supplies to meet human needs, energy efficiency and conservation measures, public health concerns and environmental protection (at all scales, from the biosphere to the local level). After investigating the resource, economic, environmental and safety aspects of fossil fuels, nuclear energy, woodfuel and other renewable sources of energy, and contemplating issues of energy efficiency and energy conservation, the Commission concludes that a 'safe, environmentally sound, and economically viable energy pathway that will sustain human progress into the distant future is clearly imperative' (WCED 1987, p. 202).

In 1992, the result of the UN Conference on Environment and Development was the adoption of the global programme, entitled Agenda 21 and the Rio Declaration on Environment and Development. Both documents identify actions to be taken to achieve the objectives of sustainable development, although neither mentions energy issues explicitly. Principle 5 of the Rio Declaration asserts that eradicating poverty is an indispensable requirement for sustainable development and that the provision of energy services is a precondition for poverty eradication. Principle 8 calls for eliminating 'unsustainable patterns of production and consumption' and has clear implications for energy production and use in developed countries with high energy intensities and in poverty-stricken developing countries. Finally, Principle 15 stipulates the wide application of the precautionary approach that is often cited in the context of climate change, where carbon dioxide (CO_2) emissions from fossil fuels are considered a possible cause of potentially serious or irreversible environmental damage.

The Millennium Summit in 2000 confirmed that progress towards sustainable development and poverty eradication has top priority for the global community.

The Millennium Development Goals (MDGs), derived from agreements and resolutions of relevant UN conferences in the post-Rio years, are rather ambitious. Some goals are only tenuously related to energy provision and use – for example, Goals 4, 5, and 6 focus on human health concerns. Other goals have important indirect implications for energy development: access to electricity could foster universal primary education (Goal 2) in many regions of the world. Improving the availability and affordability of commercial energy for cooking would drastically reduce the time and effort women need to spend collecting fuelwood, and thus promote gender equality and the empowerment of women (Goal 3).

At the macro policy level, Goal 7 calls for integrating the principles of sustainable development into country policies, using energy intensity and per capita CO_2 emissions, among others, as indicators for measuring progress. There are ample opportunities to make progress on this goal, and many economists suggest that eliminating perverse subsidies that distort the energy sector in many countries would be a good start. Ironically, some energy-related measures aimed at poverty alleviation could have negative effects on the sustainability indicators related to energy intensity or emissions in the short term. Providing electricity to promote education, increased industrialisation and urbanisation, for example, would increase the energy use per unit of gross domestic product (GDP), and replacing fuelwood with commercial fossil energy would result in higher CO_2 emissions per capita. However, once these investments in infrastructure and human capital (education, gender equality) start paying off, the energy and carbon intensity indicators should improve as well.

The World Summit on Sustainable Development (WSSD) that took place in Johannesburg in 2002 recognised that although some progress has been made, major challenges still must be overcome to implement the vision of sustainable development. Paragraph 18 of the Johannesburg Declaration on Sustainable Development lists energy among 'essential' needs and suggests rapidly increasing 'access to such basic requirements as clean water, sanitation, adequate shelter, energy, health care, food security and the protection of biodiversity' (UN 2002, p. 3). Point 9 of the Plan of Implementation of the WSSD makes a direct link between access to reliable and affordable energy services and facilitating the MDGs in general and eradicating poverty in particular. Actions to this end range from improving access to modern biomass technologies and encouraging the use of cleaner liquid fossil fuels, to developing national energy policies and regulatory frameworks and enhancing international financial and technical assistance.

Over the past 40 years the world has come a long way towards understanding the various energy–development linkages, but progress in achieving the objectives repeatedly declared in the above sources has been modest. The four basic indicators – population, GDP, primary energy consumption and CO_2 emissions – have continued to increase in most countries and at the global level (see Fig. 1). Yet these global aggregates hide a diverse picture: most of the world's population lives in developing countries, whereas the bulk of the world's GDP is generated in developed countries that also consume more energy and produce more CO_2 emissions.

Figure 2 illustrates the trends in the relationships between the four basic indicators shown in Fig. 1. Over the last four decades, the increasing global GDP per capita

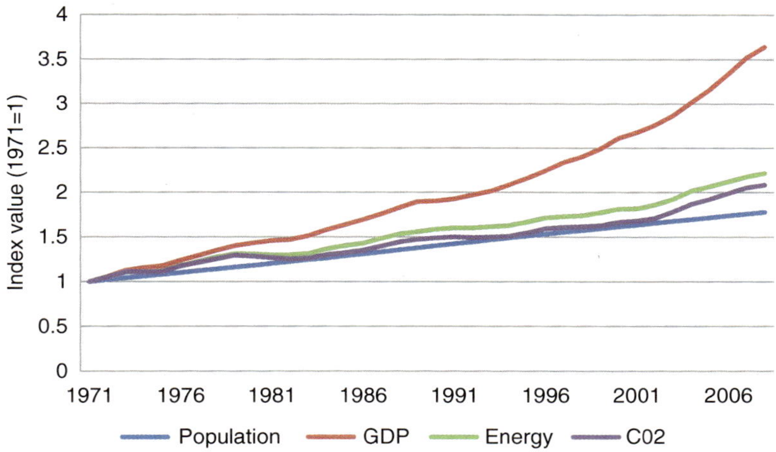

Fig. 1 Trends in global population, GDP, energy consumption and CO_2 emissions, 1971–2008 (Data source: IEA 2010a)

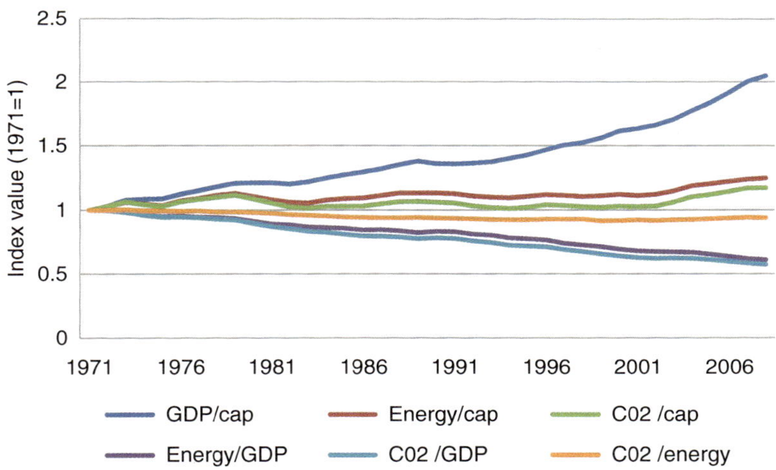

Fig. 2 Trends in per capita GDP, energy use and CO_2 emissions, energy and CO_2 intensity of GDP and the CO_2 intensity of the global energy system, 1971–2008 (Data source: IEA 2010a)

has been accompanied by declining energy use per unit of GDP, demonstrating persistent improvements in overall energy efficiency. These efficiency improvements, and the increasing share of non-fossil energy sources, have contributed to a modest decarbonisation of the energy system (CO_2/energy) and to a marked reduction in the carbon intensity of GDP.

These trends indicate that as long as rates of change in key drivers (especially GDP and to some extent population growth) exceed those of the energy and carbon intensity indicators (energy/GDP and CO_2/energy), it will not be possible to move towards sustainable energy development. Energy and increasing energy use are indispensable for development, especially in less developed countries. A transformation of the energy system will be required to achieve sustainability both on the resource side (depletable, nonrecyclable energy resources) and on the environmental side (from indoor air pollution to greenhouse gas emissions).

3 Energy–Development Linkages

Many studies have explored the relationships between economic development (GDP and its growth) and energy use (total primary or final energy, aggregated at the national level or for the main economic sectors). In addition to improving understanding of the energy system and its role in economic development, these studies are also intended to provide information for energy policy. Going beyond the widespread consensus about the existence of strong correlations between total energy consumption and its main components (fossil fuels, hydropower, other renewables and nuclear energy) and socio-economic development (usually measured as GDP growth), statistical causality analysis has become a widely used tool in the past 10 years for investigating energy–development linkages in more detail. The objective of a causality analysis is to determine whether one time series (in our case GDP) is useful for forecasting another time series (in our case energy consumption), and vice versa.

Analysts find robust evidence of co-integration but diverse directions of causal relationships between energy use and economic development. Each of the four types of causality direction (lack of causality, uni-directional causality from energy to GDP or in the opposite direction and bi-directional causality) have been detected by several studies. The causality types are as follows:

- *Neutrality hypothesis*: no causality in any direction between energy consumption and economic growth (Wolde-Rufael 2005; Jobert and Karanfil 2007; Chontanawat et al. 2008; Chiou-Wei et al. 2008).
- *Conservation hypothesis*: uni-directional causality running from GDP to energy use, which implies that energy conservation has little or no adverse effect on economic growth (Al-Iriani 2006; Yoo 2006a; Mehrara 2007; Halicioglu 2009).
- *Growth hypothesis*: uni-directional causality running from energy consumption to economic growth, which implies that limiting energy use may have a negative effect on GDP, whereas an increase in energy use would lead to economic growth (Narayan and Smyth 2008; Bowden and Payne 2009; Chang 2010; Lean and Smyth 2010).
- *Feedback hypothesis*: bi-directional causality between energy use and GDP, which implies that energy consumption and output are jointly determined and affected at the same time, therefore, any limitation on energy use would have a negative effect on economic growth. Likewise, any increase in GDP will lead to an increase in energy demand (Soytas and Sari 2006; Asafu-Adjaye and Mahadevan 2007; Apergis and Payne 2009a; Warr and Ayres 2010).

Box 1 The Granger Causality Test

The Granger causality test is a time series technique in which current and lagged values of energy consumption and economic growth are used to test for predictive power. In the Granger sense, 'X causes Y' means that 'X contains useful information for predicting Y' (Diebold 2004). The direction of the causal relationship between the two variables depends on the respective *t*-test significance when each one is regressed against the other.

More specifically, to test the hypothesis that *energy* Granger-causes *GDP*, the autoregression of GDP is augmented by including the proper lagged values of GDP and energy. If lagged values of *energy* are individually significant (according to their *t*-statistics in the regression), energy causes GDP (growth hypothesis). Conversely, to test the hypothesis that *GDP* Granger-causes *energy*, the autoregression of energy is augmented accordingly, and if the lagged values of GDP are individually significant, GDP causes energy (conservative hypothesis). Obviously, when both tests prove to be significant, the feedback hypothesis is accepted; whereas in the opposite case, when none of the tests is significant, the neutrality hypothesis holds.

Many causality studies also search for short-run and long-run causality that indicate the presence of causality over a short (few time steps) or a long (many time steps) time horizon. This analysis requires the use of vector error correction models (VECMs), depending on the approach for determining causality. Long-run causality is determined by the error correction term (if significant, long-run causality from X to Y). Short-run causality is determined with a test on the joint significance of the lagged explanatory variables, using an F-test or Wald test.

Table 1 summarises the main features and conclusions of recent statistical causality assessments of energy–economic development linkages. All studies included in the table are based on the Granger causality test (see Box 1). The literature shows no consistent empirical findings on causality directions. The outcomes are widely mixed globally, by region, by nation and within nations.

As mentioned above, each of the four causality directions are associated with different guidelines for policy making. Empirical evidence of causality running from energy use to GDP (growth hypothesis) or bi-directional causality (feedback hypothesis) indicates the energy dependency of the economy and reveals the drastic effects an energy crisis or energy conservation policies would impose on the economy. As any increase in energy consumption directly affects economic growth, and economic growth stimulates further energy consumption, these contexts are likely to be more supportive of policies that prevent energy shortages and keep up with the levels of energy consumption in order not to damage the economy.

Table 1 Assessments of statistical causality between energy and economic development

Country/Region	Source	Period	Methods/Tests	Data	Variables	Hypothesis results
12 Commonwealth of Independent States	Apergis and Payne (2009b)	1991–2005	ADF Pedroni VECM	PD	Real GDP, energy consumption, real gross fixed capital formation, and labour force	Growth (SR) Feedback (LR)
United States	Bowden and Payne (2009)	1949–2006	ADF PP Toda–Yamamoto	TS	Real GDP, industrial primary energy consumption and commercial and residential primary energy consumption	Growth (industrial primary energy consumption) Feedback (commercial and residential primary energy consumption)
United States	Warr and Ayres (2010)	1946–2000	PP Johansen VECM	TS	Exergy and GDP	Growth (SR) Feedback (LR)
United States, Thailand, Republic of Korea	Chiou-Wei et al. (2008)	1954–2006	ADF Johansen VECM G-causality	TS	Real GDP and energy consumption	Neutrality
Philippines and Singapore						Conservative
Taiwan, China; HK, Malaysia and Indonesia						Growth
22 developed countries	Lee and Chang (2007)	1965–2002	PP KPSS VAR	TS	Energy consumption per capita and real GDP per capita	Feedback
18 developing countries						Growth
France	Ang (2007)	1960–2000	VECM	TS	GDP, pollutant emissions and energy consumption	Growth (SR) Conservative (LR)
GCC countries	Al-Iriani (2006)	1971–2002	ADF Johansen Pedroni	PD	GDP and energy consumption	Conservative
11 oil exporters	Mehrara (2007)	1971–2002	ADF PP Pedroni	PD	Per capita GDP and per capita energy consumption	Conservative

G7 countries	Narayan and Smyth (2008)	1972–2002	ADF Pedroni Westerlund G-causality	PD	Capital formation, energy consumption and real GDP	Growth (LR)
G7 countries Canada, Italy, Japan, UK US and France Germany	Soytas and Sari (2006)	1960–2004	Johansen VECM	TS	Energy consumption and GDP	Feedback Growth Conservative
G7 countries United States UK Japan	Zachariadis (2007)	1960–2004	VECM, ARDL Toda–Yamamoto	TS	Real GDP and energy use	Neutrality (for the 3 models) Conservative Feedback
25 OECD countries	Belke et al. (2011)	1981–2007	ADF PP Johansen DOLS VECM	PD	Final energy consumption per capita, real GDP per capita in constant 2000 US dollars (PPP) and energy price index in US dollars (PPP)	Feedback
OECD and non-OECD countries Japan, China India, Republic of Korea and South Africa	Jinke et al. (2008)	1980–2005	ADF VECM	TS	Coal consumption and GDP	Conservative Neutrality
30 OECD countries 78 non-OECD countries China, India	Chontanawat et al. (2008)	1971–2000	ADF Johansen Hsiao	TS	Real GDP per capita and energy consumption per capita	Growth (overall) Neutrality (overall) Neutrality
China	Chang (2010)		PP Johansen VECM	PD	CO_2 emissions, energy consumption and real GDP	Growth
China	Shiu and Lam (2004) Yuan et al. (2007)	1971–2000	ECM VECM	TS	Electricity consumption and real GDP	Growth

(continued)

Table 1 (continued)

Country/Region	Source	Period	Methods/Tests	Data	Variables	Hypothesis results
China	Han et al. (2004)	1978–2000	ADF G-causality	TS	Energy consumption and real GDP	Feedback
Taiwan, China	Lee and Chang (2005)	1954–2003	ADF PP Johansen	TS	Real GDP and energy consumption (coal, oil, gas, and electricity)	Growth (LR)
Taiwan, China	Yang (2000)	1954–1997	PP Hsiao G-causality	TS	GDP and energy consumption (coal, oil, natural gas, and electricity)	Feedback
Republic of Korea	Oh and Lee (2004)	1981–2000	VECM	TS	Energy, GDP, real energy price, capital, and labour	Neutrality (SR) Conservative (LR)
Republic of Korea	Yoo (2006b)	1968–2002	ECM	TS	Coal consumption and economic growth	Feedback
Indonesia and Thailand Malaysia, Singapore	Yoo (2006a)	1971–2002	PP E–G G-causality Hsiao	TS	Real GDP per capita and electricity consumption per capita	Conservative Feedback
India, Indonesia Philippines, Thailand	Asafu-Adjaye (2000)	1973-1995	ADF PP ECM	TS	Commercial energy use per capita, GDP in constant prices and price (CPI)	Growth (SR) Feedback (SR)
India	Paul and Bhattacharya (2004)	1950–1996	E–G G-causality Johansen VECM	TS	Energy consumption and GDP	Feedback
5 ASEAN	Lean and Smyth (2010)	1980–2006	G-causality VECM	PD	Carbon dioxide emissions, electricity consumption and real GDP	Growth (LR)

Bangladesh	Mozumder and Marathe (2007)	1971–1999	ADF Johansen G-causality VECM	TS	Per capita electricity consumption and per capita GDP	Conservative
Turkey	Erdal et al. (2008)	1950–2000	ADF PP G-causality Johansen	TS	Primary energy consumption and real GNP	Growth
Turkey	Jobert and Karanfil (2007)	1960–2002	ADF PP Johansen VAR	TS	Energy consumption, energy consumption per capita, income and income per capita (real GNP)	Neutrality
Turkey	Halicioglu (2009)	1960–2005	ADF PP ARDL VECM	TS	Carbon emissions, energy consumption, income, and foreign trade	Conservative
Romania	Kayhan et al. (2010)	2001–2010	ADF PP G-causality Toda- Yamamoto	TS	Electricity consumption and GDP	Growth
Latin-America Brazil, Peru, Uruguay	Chang and Soruco Carballo (2011)	1971–2000	PP VAR VECM	TS	Energy consumption, economic growth and carbon dioxide emissions	Growth
Argentina, DRP, Mexico, Panama						Conservative
Central American countries	Apergis and Payne (2009a)	1971–2004	ADF E-G Pedroni VECM	PD	Real GDP, energy consumption, real gross fixed capital formation, and labour force	Feedback (LR)
Caribbean countries Haiti, Jamaica, Trinidad & Tobago	Francis et al. (2007)	1971–2002	ADF PP BVAR	TS	Energy consumption per capita and real GDP per capita	Feedback (SR)

(continued)

Table 1 (continued)

Country/Region	Source	Period	Methods/Tests	Data	Variables	Hypothesis results
9 Pacific Island	Mishra et al. (2009)	1980–2005	ADF PP Pedroni	PD	Energy consumption, urbanisation and GDP	Feedback
19 African countries	Wolde-Rufael (2005)	1971–2001	Toda–Yamamoto VECM	TS	Energy use per capita and per capita real GDP	Neutrality Conservative Growth Feedback
Côte d'Ivoire	Kouakou (2011)	1971–2008	ADF PP ARDL VECM	TS	Electric power industry and the economic growth	Feedback (SR) Growth (LR)
Tunisia	Belloumi (2009)	1971–2004	ADFPP Johansen VECM	TS	Per capita energy consumption and per capita GDP	Growth (SR) Feedback (LR)
Croatia	Tomislav (2009)	1953–2005	ADF PP KPSS Johansen VAR G-causality	TS	GDP and total primary energy consumption	Growth
Energy exporters	Asafu-Adjaye and Mahadevan (2007)	1971–2002	Johansen Pedroni VECM	PD	Energy consumption and real per capita GDP (constant 2000 US$) (as proxy for economic growth)	Feedback (SR, LR)

Notes: *TS* time series, *PD* panel data, *SR* short run, *LR* long run
ADF: Augmented Dickey–Fuller is a test for a unit root in a time series sample and serves to test the null hypothesis of no cointegration. The more negative the ADF statistic is, the stronger the rejection of the hypothesis that there is a unit root at some level of confidence
Pedroni: a panel cointegration test that takes the null hypothesis of no cointegration and allows heterogeneity among individual members of the panel in both the long-run cointegrating vectors and in the dynamics, since there is no reason to believe that all parameters are the same across countries

VECM: a vector error correction model, by adding error correction features to a vector auto regression model, can lead to a better understanding of the nature of any non-stationarity among different series and improve longer term forecasting

PP: the Phillips–Perron test for unit root in time series analysis tests the null hypothesis that a time series is integrated of order 1. Advantage of the PP tests over the ADF tests is that it allows for autocorrelated residuals and the user does not have to specify a lag length for the test regression

Toda and Yamamoto: an alternative causality test based upon the test equations of Granger valid for integrated or cointegrated variables. It augments extra lags depending on the potential order of integration of the series of interest. If both variables are assumed I(0), no extra lag is added in the equation, and the test is equivalent to the Granger causality test. If the series are assumed I(1), one extra lag is added to each variable in the test equation and Granger causality is conducted easier as there is no need to test cointegration or transform VAR into VECM

Johansen: a cointegration test; its desirable properties are: it rules out cointegrating relationships between the sequences and treats all variables as endogenous

KPSS: Kwiatkowski–Phillips–Schmidt–Shin stationary test a null hypothesis that an observable time series is stationary around a deterministic trend

VAR: the vector auto regression model describes the evolution of a set of k variables (called endogenous variables) over the same sample period ($t = 1 \ldots T$) as a linear function of only their past evolution. When the series are non-stationary in levels and are not cointegrated, in order to test for causality a VAR model is used where the series are first differenced

Westerlund: is a panel cointegration test which takes cointegration as the null and accommodates multiple structural breaks in both the level and trend. The test is general enough to allow for endogenous regressors, serial correlations and an unknown number of breaks that may be located at different dates for different individuals

ARDL: Autoregressive distributed lag cointegration approach has numerous advantages in comparison with other cointegration methods such as Engle and Granger and Johansen First, it can be applied irrespective of whether the regressors are purely I(0), I(1) or mutually cointegrated. Second, is the more statistically significant approach to determine cointegration in small samples. Third, allows that variables have different optimal lags, which is impossible with conventional cointegration procedures

DOLS: Dynamic ordinary least squares estimator corrects standard OLS for bias induced by endogeneity and serial correlation

Hsiao: a causality test; it is a variant of the Granger causality test that overcomes the arbitration lag selection at the heart of the Granger test

G-causality: a time series X is said to Granger-cause Y if it can be shown, usually through a series of *t*-tests and F-tests on lagged values of X (and with lagged values of Y also included), that those X values provide statistically significant information about future values of Y

E-G: Engle-Granger test is a 2 step procedure for cointegration analysis to test the null hypothesis of no cointegration between a set of I(1) variables. It estimates the coefficients of a static relationship between these variables by OLS and applies well-known unit root tests to the residuals. More recently, it has become possible to test the null hypothesis that the estimated linear relationship between the I(1) variables is a cointegrating relationship (errors in the regression are stationary)

BVAR: Bayesian VAR difference with standard VAR models relies on the fact that the model parameters are treated as random variables, and prior beliefs in form of probabilities are assigned to them

On the contrary, empirical evidence of causality running from GDP to energy (conservative hypothesis) implies that energy conservation policies will cause no significant harm to the economy; therefore, it would support the promotion of energy-efficiency practices. Moreover, results of causality studies concluding that higher GDP does not cause higher energy consumption, or that there is no causal relationship between the two (neutrality hypothesis), imply that higher energy consumption does not lead to faster economic growth, therefore, energy conservation policies could be implemented without impeding economic growth.

Table 1 also shows that causality results often vary even for the same country, sample period or similar datasets, leaving the differences in the estimation methods applied as the main cause of the divergent results. Mishra et al. (2009) have found additional methodological limitations to reaching a consensus in the literature caused by the so-called omitted variables bias in bivariate models. They suggest that, in addition to energy consumption and GDP, multivariate models should also include variables such as capital, labour and energy prices. Moreover, aggregate level studies provide poor information about causality and thus do not provide a firm/suitable basis for policy recommendations as different industrial sectors have diverse energy intensities.

The nature and direction of causality also depend on the achieved level of economic development. According to Lee and Chang (2007), increasing GDP per capita leads to energy consumption growth in developing countries, while in developed countries there is a feedback between the two. Huang et al. (2008) find no causal relationship between energy consumption and economic growth in low-income countries while economic growth affects energy consumption positively in middle-income countries and negatively in high-income countries.

The Human Development Index (HDI) (UNDP 2010) goes beyond GDP and is a more compelling socio-economic indicator that captures different aspects of human needs. The few empirical studies of the linkages between energy use and HDI to date indicate a strong correlation between energy consumption per capita and improved living standards (Goldemberg 2001; Smil 2003; Dias et al. 2006). Some studies find strong correlations between energy use and/or carbon emissions and living standards (measured as HDI) in the developing world, and decoupling at higher consumption levels (Pasternak 2000; Martinez and Ebenhack 2008). In contrast, another group of studies find no evidence of a turning point in per capita carbon emissions beyond a certain level of per capita GDP (Shafik 1994; Sharker et al. 2010). Some studies find that carbon emissions are not related to GDP per capita (Lantz and Feng 2006; Lean and Smyth 2010) while others conclude that a positive correlation between income and carbon emissions persists even for very high income levels (Arrow et al. 1995). Indeed, evidence for the Environmental Kuznets Curve (EKC) hypothesis for carbon emissions is ‘at best mixed’ (Galeotti et al. 2006) and largely inconclusive (Grubb et al. 2006).

The implication of the first group of studies is that high levels of energy use are not a precondition for attaining high GDP levels, which are therefore achievable at lower levels of energy consumption and CO_2 emissions (Steinberger and Roberts 2010). Furthermore, Martinez and Ebenhack (2008) find that, when the consumption patterns of certain 'special-cases' (a few countries with unique energy–GDP patterns) are isolated, the correlation is stronger, suggesting that large increases in human welfare (measured as HDI) are possible by relatively small increases of energy use in less developed countries.

Obviously, the second set of studies lead to opposite conclusions. They find that as income grows, carbon emissions steadily increase and no saturation point can be identified beyond which increasing GDP does not generate higher carbon emissions.

At present, a significant number of countries lie below the proposed lowest energy supply threshold of 100 kWh per capita of electricity and 100 kg of oil equivalent (kgoe) of modern fuels per capita per year (IEA 2009). Martinez and Ebenhack (2008) estimate that increasing the energy supply by some 400 kgoe per capita from modern energy sources for people in the poorest nations (with current HDI values less than 0.4) could foster a doubling of HDI.

4 Issues Concerning Energy for Development

When discussing the issue 'Energy for Development' it is important to set the stage by identifying the key energy issues that hamper development and offer possible ways to overcome them. Energy poverty and the lack of access to modern energy services like electricity, the quality and affordability of such services to those who have access, and energy insecurity are a few examples. Although energy is not explicitly included among the Millennium Development Goals, there is general agreement that eradicating energy poverty is an essential prerequisite for achieving them. This will require working towards bold targets like electricity and cleaner cooking and heating systems for all by 2030 (Brew-Hammond 2012).

An important issue for framing the energy predicament is climate change. Developing countries are most vulnerable and their best defence against the impacts will be development. Their contribution to efforts to reduce greenhouse gas emissions is important for a successful global climate policy but the financial burden should be carried by developed countries. This will require at least three international institutions with the following functions: mobilising contributions from developed countries, arranging distribution among developing countries and channelling the resources, and monitoring accountability. The estimated resources required would be on the order of 1% of GDP of developed countries (Schelling 2012).

Within the international community, including the UN system, work on energy is currently fragmented and this might be alleviated by establishing a global energy organisation. Such an organisation could help its member states in energy planning, technology transfer, in research and development on technologies to meet the energy needs of the poor countries and in ensuring energy supplies in crisis situations.

Considering the apparent relationship between income poverty and energy poverty, the triple challenge is to promote rapid economic growth, which will require equally rapid growth in energy supply and which, in turn, will need to come from clean, low-carbon sources. Several international forums are already working on various aspects, including UN-Energy, and the initiatives of the UN Secretary General involving private energy companies. But there is still need for a new institution to address the overall energy issues in a much broader context than organisations with limited mandates such as the International Atomic Energy Agency (IAEA), the UN Industrial Development Organization (UNIDO), the Organization of the Petroleum Exporting Countries (OPEC), etc. (Yumkella 2012). Another possibility might be to extend the mandate and the membership of existing organisations such as the Energy Charter Treaty (Mernier 2012).

A new international energy organisation could address energy–development challenges by searching for technology solutions, elaborating policy frameworks and monitoring implementation. An important task would be to foster relevant international treaties that bring together resource-rich and resource-poor, developing and developed countries.

These international arrangements should be complemented by providing support to local energy initiatives based on local skills, technologies and material. An important element of both types of initiatives is efficiency improvements. There is a consensus that crafting robust national energy strategies is a prerequisite for reducing energy poverty and charting sustainable energy development. The planning tools and capacity building provided by the IAEA – e.g. the Model for Analysis of Energy Demand (MAED) and the Model for Energy Supply Strategy Alternatives and their General Environmental Impact (MESSAGE) – could make important contributions towards this objective, but much more is needed. At all levels of energy decision making, there is a need to integrate all drivers (from demography to infrastructure development) that shape energy demand and all factors (from resources to technologies) on the supply side. This requirement makes energy decision making a complex and strenuous process. Good energy decision making requires learning from the past, assessing the future, considering processes and implications beyond the energy sector and compromising on 'good enough' outcomes. It must be flexible and based on good data and analysis (Cirillo 2012).

How much energy is needed for development? An important task is to explore a wide range of issues that influence energy demand now and in the future in various regions of the world. In addition to economic growth, technology, policy and lifestyle choices, demographic factors (population growth or decline, ageing, household size) and urbanisation are important drivers of energy demand (Pachauri 2012). The steady increase in the share of urban population in the coming decades will increase the demand for modern energy forms, like electricity, escalated by the need to provide access to energy for the growing populations of urban slums.

Another important driver of energy demand is infrastructure. Many developing countries are in the process of establishing or expanding their basic infrastructure (from settlement patterns to roads and buildings). This is an energy-intensive process in itself and will considerably affect energy demand during the long lifetime of the infrastructure (Luna and Gomelsky 2012). Hence, there is a chance and need for public policies to build the 'right' infrastructure for development. Improvements in the efficiency of use of available energy resources are also important in determining energy demand. Pursuing energy efficiency has many benefits, ranging from the possibility to improve access to energy, enhance energy security and reduce reliance on imports, to economic gains (such as improving returns on rural investments) and health and education benefits (Kan 2012).

Energy production and use involve many negative environmental and social impacts called externalities. The external costs need to be assessed so that they can be incorporated in prices by adequate regulation, for example, pollution taxes or tradable permit systems. Despite pervasive uncertainties, the scientific basis for assessing energy-related externalities is well established but social aspects, especially ethical considerations, remain controversial (Hirschberg 2012).

The connecting theme between energy demand and supply is technology and innovation. Funding is needed for research and development (R&D) that is of key importance on both the demand and supply sides. Implementing the results of energy R&D allows mobilising increasing quantities of resources and using them more efficiently. It helps moving down the learning curve and reducing the unit cost of energy services. It can also help in increasing the size of energy system equipment and the companies using them, thereby reducing the costs even further (Gibbons 2012).

The world is not at risk of running out of any of the energy resources currently used to provide energy services in the foreseeable future. Occurrences of fossil and fissile energy resources are plentiful, but investments at the right time and in the right quantities are needed to deliver them when and where they are needed (Rogner 2012). Investments will also be needed to satisfy the steadily growing demand for energy, as projected by the International Energy Agency (IEA 2010b), for example, but enticing private capital will require institutional improvements, like reforms in the power sector. If done well, such reforms can improve the performance, service quality and thus the affordability of electricity for the poor, but in addition to the energy sector reforms, complementary measures are often needed, such as incentive schemes, subsidies or regulatory mandates to ensure equity (Difiglio 2012).

Despite the many advantages of renewable energy (improved energy security, reduced price volatility), only a small portion of the technical potential is currently used due to the challenges involved (intermittency, large land requirements). Funding is also key to improving the technologies for harnessing renewable energy sources, and innovative financing arrangements are required to increase their contribution to the energy mix (Gupta 2012). Fossil fuels will continue to provide the major share of global energy needs in the coming decades. The main benefits of cleaner fossil technologies, especially complemented with CO_2 capture and storage (CCS), include stretching available energy reserves, keeping fossil fuels in the energy mix, thereby increasing energy security, and reducing local air pollution and greenhouse gas emissions (Burnard and McCoy 2012). Despite the recent accident at the Fukushima nuclear power plant in Japan, triggered by the combination of two extraordinary natural disasters – an earthquake and a tsunami – there are good prospects that nuclear energy will play a larger role in global energy supply due to the progress being made on safety, economics, waste management and proliferation resistance in the current and future reactor technologies and fuel cycles (Bouchard 2012). A common aspect of these two technologies, the geological disposal of CO_2

and radioactive waste offers opportunities for leaning and sharing experience (see Toth 2011). Biofuels might also become an increasingly important element of energy supply but thorough assessments and appropriate policies will be required to avoid conflicts with food security targets due to the increasing competition for land and associated higher agricultural prices (Hasan 2012).

As demonstrated by the above highlights of the topics addressed in this volume, the issues encompassed in the energy–development theme are diverse and there are rich interconnections among them. Accordingly, all of the chapters discuss topics beyond their main theme. Table 2 presents an overview of the main and secondary themes discussed in the chapters in this book.

5 Concluding Remarks

Several chapters in this volume and various statistical studies reviewed in Sect. 3 above, indicate that the HDI and energy use are closely related. At lower levels of development, a small increase in per capita energy consumption can produce a relatively large increase in a country's HDI. In developed countries, growth in energy demand in the future will be small but in the developing world massive increases in energy supply will be required to significantly improve access to energy. In some countries annual growth rates of 9–10% in electricity production are not unusual and would be desirable in many others. It is important to emphasise that a growth rate of 10% does not mean that electricity use per capita is increasing, but that more people are connected to the electricity grid. Thus every 10% increase in electricity production essentially means almost 10% additional people with access to electricity. Therefore this growth rate in electricity demand will be sustained for at least the next several decades and will only decline when a certain level of electricity consumption and a basic standard of living are attained.

It is important to recognise the contrast in terms of where the motivation comes from to improve access to energy in the developing world: it is the concern about climate change mostly in developed countries on the one hand, and the thinking about the way the developing countries should develop their energy resources on the other. The latter is focused on development targets like the MDGs (UN 2000): what kind of progress is achieved in eliminating poverty, improving health and education, and making the benefits of electricity available to the people who need it, as well as providing access to clean water. These two motivations involve different approaches to energy for development. The related questions include what can be done to help eliminate poverty, to bring electricity to billions of people who do not have it, and to prevent atmospheric pollution at the global and household levels? Millions of households in developing countries use wood for cooking, and the resulting indoor air pollution is responsible for more than 1.5 million deaths each year (WHO 2006, 2007). The answers to these questions, in turn, should also consider global implications like climate change.

There is broad agreement that only continued development will reduce the vulnerability of the poor countries to impacts of climate change, vector-borne

Table 2 Thematic coverage of the chapters in this volume

Chapter	Poverty	MGDs	Energy access	Energy policy/DM	Development institutions	Investments	Financing	Technologies	Resources	Efficiency	Climate change	Local environment
Schelling	○				●	○	●	○			●	
Brew-Hammond	○	●	●	○	○		○					○
Yumkella	○	○	●		●						●	○
Mernier	○		○		●	●	○			○		
Cirillo				●		●		●				
Pachauri	○		●	○		○		●				○
Luna and Gomelsky			○	●		●		○	○	○		
Kan	○			●		○	○			●	○	
Hirschberg				●		○		●			○	○
Gibbons				●		○		●	●		○	
Rogner				○		○		●	●			
Difiglio			○	○	○	●	●	○				
Gupta	○			○		○	●	●			○	
Burnard and McCoy				○		○		●	○		●	
Bouchard						○		●			●	
Hasan	○	○		○				●			●	

Notes: ● main theme; ○ secondary theme

MDGs Millennium Development Goals, *DM* decision making

diseases like malaria and others. It is therefore crucial to recognise that demands that the developing world participate in reducing the use of fossil fuels must not be allowed to slow down their development, which will be crucial if they are to be able to adapt to the impacts of climate change they may experience in the second half of this century. This gives a slightly different emphasis on the kinds of energy technologies between those who only focus on the elimination of poverty compared with those who focus on how to manage climate change over the coming century. It also follows that it will be important for the developed countries to help developing countries reduce their carbon emissions because otherwise they will simply accelerate climate change that will damage many of those people who are living on less than two dollars a day. It is crucially important to find a way, if only for the benefit of the developing world, to keep carbon emissions from continuing to climb and this will require the participation of the large developing countries.

Although accounting for the social costs of and mitigating climate change is fundamental, internalising the social and environmental costs of energy production and use is not a magic bullet. It is impossible to solve all the world's energy and environmental problems by internalising externalities, but it is important to solve at least some of them. Reducing and eliminating fossil fuel subsidies would be a good first step. Internalising the external costs of energy would very likely result in a rather different energy system than the one we have today, and would most likely increase the cost of energy. Increasing the cost of energy (ignoring health and other ancillary benefits for the moment) may make it even more difficult for some developing countries to satisfy the energy needs of their populations.

What policies would be most effective in advancing the concept of energy for development? First, at the highest level of policy making there is a need to set well-defined and comprehensive development priorities and strategies. Very often, setting clear and consistent priorities and realistic targets fails because policies are made by individual ministries, each focusing on their own mandates and areas of interest, with little coordination among them. Policies that integrate the priorities of different sectors are therefore needed. Extending energy planning and energy strategy development by including water, land and climate change issues is a possible way to integrate the priorities of different sectors in order to get the policies right (IAEA 2009).

Second, a broad range of policies is required. A legal framework is needed to develop markets. Financial policies are needed to mobilise investments on the supply side and credit policies are required to help households acquire more efficient equipment and appliances. In many countries, an important policy issue is energy sector reform, particularly in the power sector, aimed at improving governance and to make the provision of power and other forms of energy profitable in order to attract capital investment. In that way, power providers can become sources of revenue to governments rather than a drain on their budgets, and provide better energy services to citizens.

At the international level, policies that would be most effective in meeting the objective of providing energy for development are those that motivate developed countries to provide the resources needed by developing countries to reform and modernise their energy systems. This is partly a foreign policy matter that goes beyond any specific economic or financial issues, and some developed countries may be more willing than others to interact with developing countries in order to achieve that goal. Technology transfer is also very important, together with finding the most appropriate ways to transfer new energy technologies, making sure that they do not exceed the economic and manpower capacities of developing countries.

Making progress in all of these areas will require research into issues and linkages that are as yet poorly understood. The next step will be to set realistic development and energy targets, followed by the formulation and implementation of long-term strategies. The chapters in this book contribute to these issues and push the energy for development agenda, and thus make progress towards global sustainable development.

References

Al-Iriani, M. (2006). Energy–GDP relationship revisited: An example from GCC countries using panel causality. *Energy Policy, 34*(17), 3342–3350.

Ang, J. (2007). CO_2 emissions, energy consumption and output in France. *Energy Policy, 35*(10), 4772–4778.

Apergis, N., & Payne, J. (2009a). CO_2 emissions, energy usage and output in Central America. *Energy Policy, 37*(8), 3282–3286.

Apergis, N., & Payne, J. (2009b). Energy consumption and economic growth: Evidence from the Commonwealth of Independent States. *Energy Economics, 31*(5), 641–647.

Arrow, K., Bolin, B., Costanza, R., Dasgupta, P., Folke, C., et al. (1995). Economic growth, carrying capacity, and the environment. *Science, 268*, 520–521.

Asafu-Adjaye, J. (2000). The relationship between energy consumption, energy prices and economic growth: Time series evidence from Asian developing countries. *Energy Economics, 22*, 615–625.

Asafu-Adjaye, J., & Mahadevan, R. (2007). Energy consumption, economic growth and prices: A reassessment using panel VECM for developed and developing countries. *Energy Policy, 35*(4), 2481–2490.

Belke, H. A., Dreger, C., & Dobnik, F. (2011). Energy consumption and economic growth: New insights into the cointegration relationship. *Energy Economics, 33*(5), 782–789.

Belloumi, M. (2009). Energy consumption and GDP in Tunisia: Cointegration and causality analysis. *Energy Policy, 37*(7), 2745–2753.

Bouchard, J. (2012). Nuclear energy. In F. L. Toth (Ed.), *Energy for development: Resources, technologies, environment* (pp. 205–217). Dordrecht: Springer.

Bowden, N., & Payne, J. (2009). The causal relationship between U.S. energy consumption and real output: A disaggregated analysis. *Journal of Policy Modeling, 31*(2), 180–188.

Brew-Hammond, A. (2012). Energy: The missing Millennium Development Goal. In F. L. Toth (Ed.), *Energy for development: Resources, technologies, environment* (pp. 35–43). Dordrecht: Springer.

Burnard, K., & McCoy, S. (2012). Fossil fuels and carbon capture and storage. In F. L. Toth (Ed.), *Energy for development: Resources, technologies, environment* (pp. 187–203). Dordrecht: Springer.

Chang, C. C. (2010). A multivariate causality test of carbon dioxide emissions, energy consumption and economic growth in China. *Applied Energy, 87*, 3533–3537.

Chang, C.-C., & Soruco Carballo, C. F. (2011). Energy conservation and sustainable economic growth: The case of Latin America and the Caribbean. *Energy Policy, 39*(7), 4215–4221.

Chiou-Wei, S. Z., Chen, C.-F., & Zhu, Z. (2008). Economic growth and energy consumption revisited. Evidence from linear and nonlinear Granger causality. *Energy Economics, 30*, 3063–3076.
Chontanawat, J., Hunt, L. C., & Pierse, R. (2008). Does energy consumption cause economic growth? Evidence from a systematic study of over 100 countries. *Journal of Policy Modeling, 30*, 209–220.
Cirillo, R. R. (2012). Resolving mismatches in energy decision making. In F. L. Toth (Ed.), *Energy for development: Resources, technologies, environment* (pp. 63–78). Dordrecht: Springer.
Dias, R. A., Mattos, C. A., & Balestieri, J. A. P. (2006). The limits of human development and the use of energy and natural resources. *Energy Policy, 34*(9), 1026–1031.
Diebold, F. (2004). *Elements of forecasting*. Cincinnati: South-Western College Publishing.
Difiglio, C. (2012). Financing power sector investments. In F. L. Toth (Ed.), *Energy for development: Resources, technologies, environment* (pp. 161–170). Dordrecht: Springer.
Erdal, G., Erdal, H., & Esengün, K. (2008). The causality between energy consumption and economic growth in Turkey. *Energy Policy, 36*(10), 3838–3842.
Francis, B. M., Moseley, L., & Iyare, S. O. (2007). Energy consumption and projected growth in selected Caribbean countries. *Energy Economics, 29*(6), 1224–1232.
Galeotti, M., Lanza, A., & Pauli, F. (2006). Reassessing the environmental Kuznets curve for CO_2 emissions: A robustness exercise. *Ecological Economics, 57*, 152–163.
Gibbons, J. H. (2012). Technology and innovation. In F. L. Toth (Ed.), *Energy for development: Resources, technologies, environment* (pp. 141–148). Dordrecht: Springer.
Goldemberg, J. (2001). *Energy and human well-being*. New York: United Nations Development Programme.
Grubb, M., Bulter, L., & Feldman, O. (2006). *Analysis of the relationship between growth in carbon dioxide emissions and growth in income.* Cambridge: Faculty of Economics, University of Cambridge.
Gupta, S. (2012). Financing renewable energy. In F. L. Toth (Ed.), *Energy for development: Resources, technologies, environment* (pp. 171–186). Dordrecht: Springer.
Halicioglu, F. (2009). An econometric study of CO_2 emissions, energy consumption, income and foreign trade in Turkey. *Energy Policy, 37*(3), 1156–1164.
Han, Z. Y., Wei, Y. M., Jiao, J. L., Fan, Y., & Zhang, J. T. (2004). On the cointegration and causality between Chinese GDP and energy consumption. *System Engineering, 22*(12), 17–21.
Hasan, F. (2012). Energy and food security. In F. L. Toth (Ed.), *Energy for development: Resources, technologies, environment* (pp. 219–233). Dordrecht: Springer.
Hirschberg, S. (2012). Externalities in the global energy system. In F. L. Toth (Ed.), *Energy for development: Resources, technologies, environment* (pp. 121–138). Dordrecht: Springer.
Huang, B., Hwang, M. J., & Yang, C. W. (2008). Causal relationship between energy consumption and GDP growth revisited: A dynamic panel data approach. *Ecological Economics, 67*, 41–54.
IAEA (International Atomic Energy Agency). (2009). *Seeking sustainable climate, land, energy and water (CLEW) strategies*. Vienna: IAEA. www.iaea.org/About/Policy/GC/GC53/GC53InfDocuments/English/gc53inf-3-att6_en.pdf. Accessed 20 Aug 2011.
IEA (International Energy Agency). (2009). *World energy outlook 2009*. Paris: Organisation for Economic Co-operation and Development (OECD)/IEA. www.iea.org/textbase/nppdf/free/2009/weo2009.pdf
IEA (International Energy Agency). (2010a). "World Indicators", *IEA world energy Statistics and Balances* (database). doi: 10.1787/data-00514-en
IEA (International Energy Agency). (2010b). *Energy technology perspectives 2010. Scenarios & strategies to 2050*. Paris: IEA. www.iea.org/papers/2009/ETP_2010_flyer.pdf
Jinke, L., Hualing, S., & Dianming, G. (2008). Causality relationship between coal consumption and GDP: Difference of major OECD and non-OECD countries. *Applied Energy, 85*(6), 421–429.
Jobert, T., & Karanfil, F. (2007). Sectoral energy consumption by source and economic growth in Turkey. *Energy Policy, 35*(11), 5447–5456.
Kan, C.-C. F. (2012). Energy efficiency for development. In F. L. Toth (Ed.), *Energy for development: Resources, technologies, environment* (pp. 113–120). Dordrecht: Springer.

Kayhan, S., Adiguzel, U., Bayat, T., & Lebe, F. (2010). Causality relationship between real GDP and electricity consumption in Romania (2001–2010). *Journal for Economic Forecasting, 4*, 169–183.
Kouakou, A. K. (2011). Economic growth and electricity consumption in Côte d'Ivoire: Evidence from time series analysis. *Energy Policy, 39*(6), 3638–3644.
Lantz, V., & Feng, Q. (2006). Assessing income, population, and technology impacts on CO_2 emissions in Canada: Where's the EKC? *Ecological Economics, 57*, 229–238.
Lean, H. H., & Smyth, R. (2010). CO_2 emissions, electricity consumption and output in ASEAN. *Applied Energy, 87*(6), 1858–1864.
Lee, C. C., & Chang, C. P. (2005). Structural breaks, energy consumption, and economic growth revisited evidence from Taiwan. *Energy Economics, 27*(6), 857–872.
Lee, C. C., & Chang, C. P. (2007). Energy consumption and GDP revisited a panel analysis of developed and developing countries. *Energy Economics, 29*(6), 1206–1223.
Luna, N., & Gomelsky, R. (2012). Development, infrastructure and energy: Exploring the linkages in Latin America. In F. L. Toth (Ed.), *Energy for development: Resources, technologies, environment* (pp. 95–112). Dordrecht: Springer.
Martinez, D. M., & Ebenhack, B. W. (2008). Understanding the role of energy consumption in human development through the use of saturation phenomena. *Energy Policy, 36*(4), 1430–1435.
Mehrara, M. (2007). Energy consumption and economic growth: the case of oil exporting countries. *Energy Policy, 35*(5), 2939–2945.
Mernier, A. (2012). The possible role of the energy charter in advancing energy for development in Africa. In F. L. Toth (Ed.), *Energy for development: Resources, technologies, environment* (pp. 57–62). Dordrecht: Springer.
Mishra, V., Sharma, S., & Smyth, R. (2009). The energy–GDP nexus: Evidence from a panel of Pacific Island countries. *Resource and Energy Economics, 31*(3), 210–220.
Mozumder, P. P., & Marathe, A. A. (2007). Causality relationship between electricity consumption and GDP in Bangladesh. *Energy Policy, 35*, 395–402.
Narayan, P., & Smyth, R. (2008). Energy consumption and real GDP in G7 countries: New evidence from panel co-integration with structural breaks. *Energy Economics, 30*(5), 2331–2341.
Oh, W., & Lee, K. (2004). Energy consumption and economic growth in Korea testing the causality relation. *Journal of Policy Modeling, 26*, 973–981.
Pachauri, S. (2012). Demography, urbanization and energy demand. In F. L. Toth (Ed.), *Energy for development: Resources, technologies, environment* (pp. 81–94). Dordrecht: Springer.
Pasternak, A. (2000). *Global energy futures and human development: A framework for Analysis* (US Department of Energy Report UCRL-ID-140773). Livermore: Lawrence Livermore National Laboratory.
Paul, S., & Bhattacharya, R. N. (2004). Causality between energy consumption and economic growth in India: A note on conflicting results. *Energy Economics, 26*(6), 977–983.
Rogner, H.-H. (2012). Energy resources. In F. L. Toth (Ed.), *Energy for development: Resources, technologies, environment* (pp. 149–160). Dordrecht: Springer.
Schelling, T. C. (2012). Rethinking energy aid mechanisms: Three premises. In F. L. Toth (Ed.), *Energy for development: Resources, technologies, environment* (pp. 27–34). Dordrecht: Springer.
Shafik, N. (1994). Economic development and environmental quality: An econometric analysis. *Oxford Economic Papers, 46*, 757–773.
Sharker, M. A. Y., Nasser, M., & Razzaque, M. A. (2010, January 1–2). *Examining environmental Kuznets curve for carbon dioxide emission in Bangladesh: A bootstrap approach.* In Proceedings of 3rd annual Asian business research conference held at the BIAM Foundation in Dhaka, Bangladesh. Melbourne: World Business Institute.
Shiu, A., & Lam, P. L. (2004). Electricity consumption and economic growth in China. *Energy Policy, 32*, 47–54.
Smil, V. (2003). *Energy at the crossroads: Global perspectives and uncertainties*. Cambridge: MIT Press.
Soytas, U., & Sari, R. (2006). Energy consumption and income in G7 countries. *Journal of Policy Modeling, 28*, 739–750.

Steinberger, J. K., & Roberts, J. T. (2010). From constraint to sufficiency: The decoupling of energy and carbon from human needs, 1975–2005. *Ecological Economics (special section: Ecological distribution conflicts), 70*(2), 425–433.

Tomislav, G. (2009). Causality between economic growth and energy consumption in Croatia. *Journal of Economics and Business, 27*(2), 327–348.

Toth, F. L. (Ed.). (2011). *Geological disposal of carbon dioxide and radioactive waste: A comparative assessment*. Dordrecht: Springer.

UN (United Nations). (2000, September 8). *United Nations Millennium Declaration*. Resolution 2, Session 55. New York: UN. www.undemocracy.com/A-RES-55-2

UN (United Nations). (2002). *Report of the World Summit on Sustainable Development* (A/CONF.199/20). New York: UN.

UNCHE (United Nations Conference on the Human Environment). (1972). *Declaration of the United Nations Conference on the Human Environment*. New York: UN. Available from: www.unep.org/Documents.multilingual/Default.asp

UNDP (United Nations Development Program). (2010). *Human development report 2010. The real wealth of nations: Pathways to human development*. New York: UNDP. http://hdr.undp.org/en/media/HDR_2010_EN_Complete_reprint.pdf

Warr, B. S., & Ayres, R. U. (2010). Evidence of causality between the quantity and quality of energy consumption and economic growth. *Energy, 35*(4), 1688–1693.

WCED (World Commission on Environment and Development). (1987). *Our common future* (Report of the World Commission on Environment and Development). New York: Oxford University Press.

WHO (World Health Organization). (2006). *Fuel for life: Household energy and health*. Geneva: WHO. www.who.int/indoorair/publications/fuelforlife/en/index.html. Accessed 29 Apr 2010.

WHO (World Health Organization) (2007). *Indoor air pollution: National burden of disease estimates*. Geneva: WHO. www.who.int/indoorair/publications/indoor_air_national_burden_estimate_revised.pdf. Accessed 29 Apr 2010.

Wolde-Rufael, Y. (2005). Energy demand and economic growth: The African experience. *Journal of Policy Modeling, 27*(8), 891–903.

Yang, H. Y. (2000). A note on the causal relationship between energy and GDP in Taiwan. *Energy Economics, 22*, 309–317.

Yoo, S.-H. (2006a). The causal relationship between electricity consumption and economic growth in the ASEAN countries. *Energy Policy, 34*(18), 3573–3582.

Yoo, S.-H. (2006b). Causal relationship between coal consumption and economic growth in Korea. *Applied Energy, 83*(11), 1181–1189.

Yuan, J. H., Zhao, C. H., Yu, S. K., & Hu, Z.-G. (2007). Electricity consumption and economic growth in China: Cointegration and co-feature analysis. *Energy Economics, 29*, 1179–1191.

Yumkella, K. K. (2012). Multilateralism and energy for development. In F. L. Toth (Ed.), *Energy for development: Resources, technologies, environment* (pp. 45–56). Dordrecht: Springer.

Zachariadis, T. (2007). Exploring the relationship between energy use and economic growth with bivariate models: New evidence from G-7 countries. *Energy Economics, 29*(6), 1233–1253.

Part I
International Institutions and National Decisions

Rethinking Energy Aid Mechanisms: Three Premises

Thomas C. Schelling

Abstract This chapter discusses three key premises that characterise the current debate on climate change and the state of the world's energy economy. First, the developing world is far more susceptible to climate change than the developed world. Second, the best defence for developing nations against climate change will be their own economic and social advancement. Third, the rich countries will need to provide financial support to help the developing world reduce greenhouse gas emissions without disrupting or interfering substantially with their own economies. New global institutions will be needed to mobilise rich countries to provide vital energy aid to poor countries, to represent the larger developing countries to help them decide how to share those resources, and to serve as an effective intermediary to channel the resources and monitor how they are distributed and used.

1 Introduction

Three key premises serve to characterise the current debate on climate change and the state of the world's energy economy. The socio-economic and political implications of these premises are discussed in detail in this chapter.

The first premise is that the developing world is far more susceptible to damage due to climate change than the developed world. The United States appears to be far less vulnerable to climate change – unless one of the more disastrous scenarios occurs (US NAS 2008). Most market-oriented activities in the United States are performed indoors. Open heart surgery, banking and insurance, and radio

T.C. Schelling (✉)
School of Public Policy, University of Maryland, 2101 Van Munching Hall,
College Park, MD 20742, USA

F.L. Toth (ed.), *Energy for Development: Resources, Technologies, Environment*, Environment & Policy 54, DOI 10.1007/978-94-007-4162-1_2,

broadcasting can be done in almost any climate. Most manufacturing may be undertaken in almost any climate. Even much of US farming has moved indoors. Yet the country's vulnerability to climate change is largely in agriculture – which is a very small part of the US economy. This reality appears to be true in Austria, Canada and in most of the developed world.

The economic reality in the developing world is very different: up to half of the population may depend on agriculture for its existence (IPCC 2007). For that reason alone, the developing world is most at risk from climate change. Ten years ago, most spokespeople for the developing world were reluctant to admit this stark reality, but no longer. One key example is Beijing's Museum of Science, which has devoted the entire ground floor to the problem of global warming and climate change, with photographs of melting glaciers and other dramatic and threatening natural events.

A second key premise is that the best defence for developing nations against climate change will be their own economic and social advancement. Over the next 50 or 75 years the developing world should have the opportunity to continue to grow and thus become less susceptible to climate change.

The third premise is that in order for that development to happen, the richer countries will need to financially support the developing world in mitigating greenhouse gas emissions – but without disrupting or interfering substantially with their own economic development. How can the well-to-do help the poor to reduce their dependence on fossil fuels and their susceptibility to damage due to climate change?

2 Institutions

Three kinds of institution will be needed – and none of these exist at present. One institution or set of institutions must be able to mobilise the nations of the developed world to contribute economic resources to the poor countries for energy development. But no global institution is responsible for deciding which rich countries should participate, how much, in total, they should contribute, or how they would share their contributions.

More than 50 years ago, there were pioneering efforts to develop 'international cost-sharing arrangements' (Schelling 1958). Even further back, there had been efforts to allocate costs among nations beginning with the Universal Postal Union in 1874. The International Court of Justice emulated the model of the Postal Union, and so did the League of Nations, which based contributions on geographical extent, population size and other socio-economic indicators. To finance the United Nations after the Second World War there was an effort to establish the equivalent of a kind of income tax among the nations of the world (Russell 1966; Childers 1995; McDermott 2000). The United Nations Relief and Rehabilitation Administration (UNRRA) managed to assemble some US$4 billion for the first year – roughly 1% of gross world product at the time. Negotiations led finally to a formula in which every contributing nation would contribute 1% of a year's gross domestic product (GDP). But later it turned out that most of the money was donated by the United

States, which so dominated the world economy at the end of the war that it felt obliged to do most of the financing for the UNRRA.

But we currently have no way to decide which nations should contribute to energy programmes in developing countries, how much they should contribute, what their relative shares should be, how large a total should be contemplated, and how that total might change over time. This assistance would have to be devoted to precisely the kinds of energy for development strategies as emphasised by Yumkella (2012). Eventually, additional contributions may be needed to help poor countries adapt to climate change.

One critically important part of adaptation to climate change is the storage and transport of water. One area that remains poorly understood by the scientific community is what impacts global warming will have above 3,000 m (Haeberli and Beniston 1998; Hall and Fagre 2003; Watson and Haeberli 2004; De Jong et al. 2006; López-Moreno et al. 2008). Not many people live above 3,000 m, but immense areas of the world depend on the water that accumulates above 3,000 m. In the wintertime a lot of precipitation falls as snow and remains in that form until late spring or early summer when it becomes available for irrigation. If a lot of that precipitation falls as rain instead, or if what falls as snow melts too early, then a huge new infrastructure would be required to capture the water and hold it for irrigation and transport it to where it is needed. Agriculture throughout China, India, Chile, Argentina, Peru and California depends very much on what happens above 3,000 m, and that is highly uncertain.

A second institution is needed to represent the larger developing countries in order to help them decide how they should share whatever resources become available. Again, at present there is no institution with this kind of responsibility or authority. There is no venue in which countries as disparate as China, India, Indonesia, Brazil and others can get together and cooperate in arriving at a sharing arrangement for whatever assistance becomes available.

Third, there will need to be some kind of energy agency like the World Bank that can effectively channel assistance from the richer countries to poor countries. That same agency will need to monitor the way the assistance is distributed and used, and be accountable to the sponsoring countries for what kinds of energy projects receive financing.

These three institutions currently do not exist. They will need to be developed because without them the rich countries cannot assist poor countries in coping with climate change.

3 Historical Analogues

What are the precedents? Most of them are at least half a century old. Soon after the Second World War, the United Stated embarked on the Marshall Plan for Western Europe (Hogan 1989; DeLong and Eichengreen 1991; Eichengreen 1995; Kipping and Bjarnar 1998; Agnew and Entrikin 2004). In the first year, the United States allocated about US$5 billion over a 15-month period to the participating West European recipient nations. For the second year, the Unites States said the recipient nations would have to divide the total among themselves.

The division process involved every recipient nation devoting many months to documenting whatever its claim for aid was. It included indicating what every recipient nation would need by way of imports that would have to be paid for in convertible currencies; what nations were doing to ration their use of petroleum, butter and meat; how much investment would go into restoring roads, railways and canals; how much would go into building and rebuilding; and how much was needed to replenish worn-out equipment or facilities destroyed during the war (Wood 1986; Maier 1991; Sorel and Padoan 2008). Every recipient nation spent months developing what we might call a document for its claim for a share of the Marshall Plan funds (US Congress 1948; Surrey 1948; Gimbel 1976; Wood 1986; Haas 2007).

Then at the ministerial level over a period of more than 6 weeks the nations confronted each other, negotiated, cross-examined each other and tried to arrive at an agreed sharing of about US$4 billion that was becoming available. They did not quite reach agreement, but the Secretary General of the OEEC (the Organisation for European Economic Co-operation, established in 1948, which was superseded by the OECD in 1961) and the Belgian delegate (Belgium was not asking for any money) negotiated to arrive at a proposal that was unanimously accepted for the division of US aid. It was a healthy thing that the recipients of the Marshall Plan funds were themselves able to reach agreement on how the aid should be divided up in the second year. But today, there is no organisation like the OEEC that could enable the developing nations to find a way to divide the funds received from developed countries among themselves.

The only other significant precedent for such contributions arose under the North Atlantic Treaty Organization (NATO). At the beginning, in 1951, NATO was about to reach agreement on what each NATO country would contribute by way of a military preparedness effort (Loewenstein 1954; Lord Ismay 1955; Schelling 2005). The members all had to decide how many young men they would recruit into military service, how much they would pay them, and how much real estate they would provide for manoeuvres, military housing and military pipelines. They also had to decide how much they would spend on equipment and how much they would depend on equipment from the United States.

In the end NATO went through with something like the earlier Marshall Plan negotiation to arrive at an understanding of what each nation would contribute in terms of military preparedness against possible attack from the East. They actually arrived at commitments that for the most part they met, as the years went by. But again that was 50 years ago, and no comparable international commitment procedure has occurred since then.

The nations meeting at the Bretton Woods conference that took place in the United States after the Second World War were able to arrive at an understanding of how the World Bank (first named the International Bank for Reconstruction and Development) would be endowed with the financing needed to carry out its functions (Bittermann 1971; Meier 1971; Mason and Robert 1973). The same came to be true with the International Monetary Fund (IMF) (Horsefield 1969; Southard 1980). Member countries were able to agree on how much of their own currencies and how much of convertible currencies they would contribute to these two institutions, both of which have lasted for more than 60 years. But the global community has not shared anything like that process in the last half century.

The Kyoto Protocol to the UN Framework Convention on Climate Change (UN 1998) went through the motions of negotiation, but did not really arrive at anything very substantial in terms of a solid agreement.

4 Resource Requirements and Utilisation

How much money is needed for meaningful and effective programmes that will allow developing countries to shift to cleaner energy and adapt to global warming? The UN's *World Economic and Social Survey* (UN 2009) estimated that the developing world would need between US$500 billion and US$600 billion annually from the rich nations, or around 1% of their GDP. This amount is well above previous estimates but is dwarfed by the amounts needed according to a Chinese economic analysis. Environmental economist Zou Ji, at Renmin University in Beijing (Hille and Harvey 2009), suggests that China alone will need to spend up to US$438 billion annually if its greenhouse gas emissions are to peak by 2030. Again, the amount required would be something around 1% of the GDP of all the contributing rich nations combined.

Is 1% of GDP a big number or a little number? With most of the world in deep recession in 2009, contributing 1% of GDP to the developing world appears to be a lot. But when the growth of GDP per capita in the developed world is plotted over the next 60 years, per capita income at least doubles. Contributing in perpetuity 1% of the GDP of the developed world would probably mean that the per capita income, instead of doubling in 55 or 60 years, will not do so until 56 or 57, or 61 or 62 years. From that perspective, 1% of GDP is a very small amount to be subtracted from the consumption in the developed world. A 1% of GDP contribution may appear to be an economically manageable sum, but the politics of it are likely to be very different.

How should this assistance be transferred? Could a simple financial transfer be made from the rich countries to the poor ones? That is unlikely to be politically attractive. If it is simply a matter of spending money to get the Chinese or the Indians or the Brazilians or the Indonesians to participate in some kind of project to reduce their carbon dioxide emissions, a lot of people will view this transfer as bribery and/or extortion. Probably, instead, any transfer of resources will need to take the form of specific energy-related development projects in the developing world.

What kinds of projects? Compiling a detailed list of plausible projects related to improving energy services and reducing global emissions will not be simple. A good test case is the nuclear energy sector. To promote nuclear energy in the developing world it may be essential that the important components of nuclear reactors be financed by the developed world. The manpower to produce the infrastructure and cement would not look as if it deserves to be provided from abroad. But at least the components of nuclear reactors would need to be imported to make nuclear energy financially attractive.

Another set of projects will have to involve the capture and sequestration of carbon dioxide from coal-fired power plants in China, India and numerous other countries (IPCC 2005; IEAGHG 2007, 2008). These projects will require transfers of technologies that have yet to be perfected. They will necessitate a huge amount of geological exploration to find out where and how carbon can be safely stored underground, where it will not leak out for at least 200 or 300 years. This is going to require a lot of new technology, including massive pipelines from power plants to wherever the gas is sequestered.

Yet another example is methane clathrates. In the near future the huge deposits of these compounds on the continental shelves under water are likely to look very attractive as a source of energy. All over the world, off the coasts of China, Mexico and those of most continents there are huge bodies of methane clathrates, in which the methane is trapped within a combination of crystals of both methane and water (Lee and Holder 2001; Chatti et al. 2005; Boswell 2007).

It will be very important to ensure that disturbing these deposits does not lead to eruptions or leakages of methane into the atmosphere. It will take only a very small percentage of the methane in clathrates to leak into the atmosphere to make methane a worse greenhouse gas than carbon dioxide. When Mexico or China or India decide to tap these resources, they may be in too great a hurry to develop the technology that can safely extract methane without it leaking into the atmosphere. Convincing the rich countries to finance the kind of research, development and testing necessary to tap these resources may be important to make sure that the methane, when or if it is tapped, does not leak into the atmosphere.

It is almost inevitable that when countries like the United States, the more developed parts of the European Union, Canada, Australia, New Zealand and other developed countries make contributions, most will want to do so in the form of exports. They are going to want to make financial contributions that will be spent at home. This can already be seen in the United States with its foreign aid programmes. The phrase ‘buy American’ means the country would like to contribute in the form of exports because it makes local industries and businesses happy. That is likely to be a constraint on what can be provided by the well-to-do to the poor countries.

5 Conclusions

This chapter has attempted to make it clear that the institutions necessary to support rich countries in providing vital energy aid to poor countries do not exist today. Moreover, the institutions essential to compel the largest developing countries – China, India, Indonesia, Brazil and several others – to work together and share whatever resources may become available also do not currently exist. Finally, a qualified intermediary analogous to the World Bank that could channel the resources and follow up on what is achieved with those resources does not exist either. Such an agency must be able to assure the contributors that the funds are being used in the manner that they were expected to be used. We simply do not have these global facilities available.

A critical final issue is: who is working to address this problem? The answer at this moment seems to be practically no one at the international level. Who today is working on how the rich can be gathered together to agree on contributions to the poor? And who is considering how the poor countries can be gathered together – the appropriate poor, the ones concerned with greenhouse gas emissions, and whose participation is going to be crucial and necessary over the coming half century to cope with global warming? Until these new institutions are established, there is little way that the developing world can be assisted in reducing greenhouse gas emissions without interfering with their continued growth.

References

Agnew, J. J., & Entrikin, N. (Eds.). (2004). *The Marshall plan today: Model and metaphor*. London: Routledge.

Bittermann, H. J. (1971). Negotiation of the articles of agreement of the International Bank for Reconstruction and Development. *International Lawyer, 5*(1), 59–88.

Boswell, R. (2007). Resource potential of methane hydrate coming into focus. *Journal of Petroleum Science and Engineering, 56*(1–3), 9–13.

Chatti, I., Delahaye, A., Fournaison, L., & Petitet, J.-P. (2005). Benefits and drawbacks of clathrate hydrates: A review of their areas of interest. *Energy Conversion and Management, 46*(9–10), 1333–1343.

Childers, E. (1995). Financing the United Nations: Some possible solutions. *Futures, 27*(2), 161–170.

De Jong, C., Collins, D. N., & Ranzi, R. (Eds.). (2006). *Climate and hydrology in mountain areas*. Chichester: Wiley.

DeLong, J. B., & Eichengreen, B. (1991). *The Marshall plan: History's most successful structural adjustment program* (NBER Working Paper No. W3899). Cambridge: National Bureau of Economic Research. http://ssrn.com/abstract=226738. Accessed 18 Mar 2010.

Eichengreen, B. J. (1995). *Europe's post-war recovery*. Cambridge: Cambridge University Press.

Gimbel, J. (1976). *The origins of the Marshall plan*. Stanford: Stanford University Press.

Haas, J. (2007). 60 years of Marshall Plan aid: A critical appraisal from an Austrian perspective. *Monetary Policy & the Economy, 2*, 126–139.

Haeberli, W., & Beniston, M. (1998). Climate change and its impacts on glaciers and permafrost in the Alps. *Ambio, 27*(4), 258–265.

Hall, M. H. P., & Fagre, D. B. (2003). Modeled climate-induced glacier change in Glacier National Park, 1850–2100. *Bioscience, 53*(2), 131–140.

Hille, K., & Harvey, F. (2009, September 1). China's high price for emission cuts. *Financial Times*. www.ft.com/cms/s/0/cd7466e8-971f-11de-83c5-00144feabdc0.html. Accessed 3 May 2010.

Hogan, M. J. (1989). *The Marshall plan: America, Britain, and the reconstruction of western Europe, 1947–1952*. Cambridge: Cambridge University Press.

Horsefield, J. K. (Ed.). (1969). *The International Monetary Fund 1945–1965: Twenty years of international monetary cooperation* (3 Vols.). Washington, DC: International Monetary Fund.

IEAGHG (International Energy Agency Greenhouse Gas R&D Programme). (2007). *Storing CO_2 underground*. Cheltenham: IEAGHG.

IEAGHG (International Energy Agency Greenhouse Gas R&D Programme). (2008). *Geologic storage of carbon dioxide*. Cheltenham: IEAGHG.

IPCC (Intergovernmental Panel on Climate Change). (2005). *Special report on carbon dioxide capture and storage*. B. Metz, O. Davidson, H. C. de Coninck, M. Loos, & L. A. Meyer (Eds.). Cambridge, UK: Cambridge University Press.

IPCC (Intergovernmental Panel on Climate Change). (2007). *Climate change 2007: Impacts, adaptation and vulnerability* (Contribution of Working Group II to the fourth assessment report of the IPCC). M. L. Parry, O. F. Canziani, J. P. Palutikof, P. J. van der Linden, & C. E. Hanson (Eds.). Cambridge, UK: Cambridge University Press.

Ismay, L. (1955). *NATO: The first five years, 1949–1954*. Roquencourt: North Atlantic Treaty Organization.

Kipping, M., & Bjarnar, O. (Eds.). (1998). *The Americanisation of European business: The Marshall plan and the transfer of US management models*. London: Routledge.

Lee, S.-Y., & Holder, G. D. (2001). Methane hydrates potential as a future energy source. *Fuel Processing Technology, 71*(1–3), 181–186.

Loewenstein, K. (1954). Sovereignty and international co-operation. *American Journal of International Law, 48*(2), 222–244.

López-Moreno, J. I., Beniston, M., & García-Ruiz, J. M. (2008). Environmental change and water management in the Pyrenees: Facts and future perspectives for Mediterranean mountains. *Global and Planetary Change, 61*(3–4), 300–312.

Maier, Ch. S. (Ed.). (1991). *The cold war in Europe: Era of a divided continent*. New York: Markus Wiener.

Mason, E. S., & Robert, R. E. (1973). *The World Bank since Bretton Woods*. Washington, DC: Brookings Institution Press.

McDermott, A. (2000). *The new politics of financing the UN*. New York: Palgrave Macmillan.

Meier, G. M. (1971). The Bretton Woods agreement: Twenty-five years after. *Stanford Law Review, 23*(2), 235–275.

Russell, R. B. (1966). United Nations financing and the law of the charter. *Columbia Journal of Transnational Law L, 5*, 68–95.

Schelling, T. C. (1958). International cost-sharing arrangements. In T. C. Schelling (Ed.), *International economics*. Boston: Allyn and Bacon.

Schelling, T. C. (2005). What makes greenhouse sense? *Indiana Law Review, 38*, 581–594.

Sorel, E., & Padoan, P. C. (Eds.). (2008). *The Marshall plan: Lessons learned for the 21st century*. Paris: OECD Publishing.

Southard, F. A. Jr. (1980). Evolution of the International Monetary Fund. *North Carolina Journal of International Law and Commercial Regulation, 5*(3), 425–469.

Surrey, W. S. (1948). The economic cooperation Act of 1948. *California Law Review, 36*(4), 509–557.

UN (United Nations). (1998). *Kyoto Protocol to the United Nations Framework Convention on Climate Change*. New York: United Nations.

UN (United Nations). (2009). *World economic and social survey 2009*. New York: United Nations.

US Congress. (1948). *The economic cooperation act of 1948* (Publ. No.472, 80th Congress, 2nd Session). Washington, DC: US Congress.

US NAS (US National Academy of Sciences). (2008). *Understanding and responding to Climate change*. Washington, DC: Board on Atmospheric Sciences and Climate, US National Academy of Sciences. http://dels.nas.edu/dels/rpt_briefs/climate_change_2008_final.pdf. Accessed 23 Mar 2010.

Watson, R. T., & Haeberli, W. (2004). Environmental threats, mitigation strategies and high-mountain areas. In: Royal colloquium: Mountain areas – a global resource. *Ambio Special Report, 13*, 2–10.

Wood, R. E. (1986). *From Marshall plan to debt crisis: Foreign aid and development choices in the world economy*. London: University of California Press.

Yumkella, K. K. (2012). Multilateralism and energy for development. In F. L. Toth (Ed.), *Energy for development: Resources, technologies, environment* (pp. 45–56). Dordrecht: Springer.

Energy: The Missing Millennium Development Goal

Abeeku Brew-Hammond

Abstract The Millennium Development Goals (MDGs) include eradicating extreme poverty, improving the health of women and children, gender empowerment, environmental sustainability and global partnerships. There are no energy-related targets, such as access to energy, although there is increasing evidence of relationships between energy and the MDGs and development in general. The global community has to commit to eradicating energy poverty within the next few decades. This will require financial resources, knowledge and institutional resources, and stimulating national governments to lead the process of securing energy access for their peoples.

1 Introduction

Is energy the missing Millennium Development Goal (MDG)? I confess that I am one of those who were happy when energy was left out of the MDGs (UN 2000). For a long time we in the energy community had grown used to thinking of energy by itself. We were interested in doing solar energy, wind energy, nuclear energy, animal draft power and other forms of energy, but we were doing it in isolation. However, energy is only useful if it contributes to the social and economic development of people. I believe that the MDGs, by leaving out energy, forced us to do some much-needed homework. It forced us to go back to the drawing board to look at how energy contributes to different aspects of social and economic activity.

A. Brew-Hammond (✉)
The Energy Center, Kwame Nkrumah University of Science and Technology (KNUST), Kumasi, Ghana
e-mail: abeeku@brewhammond.com

F.L. Toth (ed.), *Energy for Development: Resources, Technologies, Environment*, Environment & Policy 54, DOI 10.1007/978-94-007-4162-1_3,

This chapter draws on the work of the Global Energy Assessment (GEA) to make the case for a new goal on energy access for all by 2030. Section 2 provides an overview of the energy–development linkages. Section 3 reviews issues related to access to energy by the poor. Finally, Sect. 4 elaborates a number of strategies to improve energy access.

2 Linkages Between Energy and Development

The MDGs range from eradicating extreme poverty and improving the health of women and children, to gender empowerment, environmental sustainability and, of course, developing a global partnership for development (see Fig. 1). Inspired by the MDG framework, the energy community has reconsidered some important linkages. As a result, there is now a large body of new knowledge that provides examples and demonstrates relationships between energy and the MDGs, and development in general (Halsnæs and Garg 2011; Bazilian et al. 2010; WHO and UNDP 2009; Irvine-Halliday et al. 2008).

Unfortunately, at this point in time, there still is a risk that many countries, especially those in sub-Saharan Africa, will not achieve the MDGs. Extreme poverty is indeed rising. Some 2 billion people still live on less than US$2 per day. Figure 2 shows that the worst situations occur in sub-Saharan Africa, where the number of those living below US$1.25 per day has been increasing. The same is true in South Asia, where poverty is also increasing. But East Asia and the Pacific have demonstrated that this trend does not need to continue; there are ways of breaking away from this path of increasing poverty. Millions of people in many countries have been lifted out of extreme misery.

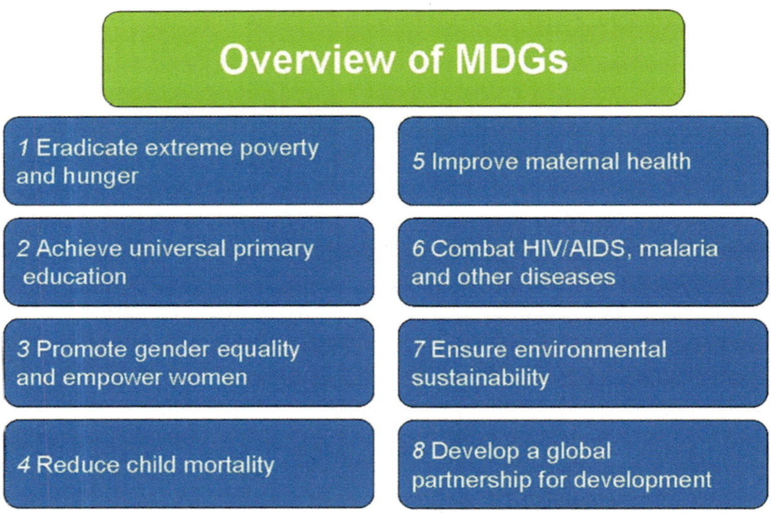

Fig. 1 Millennium Development Goals (Source: UN 2000)

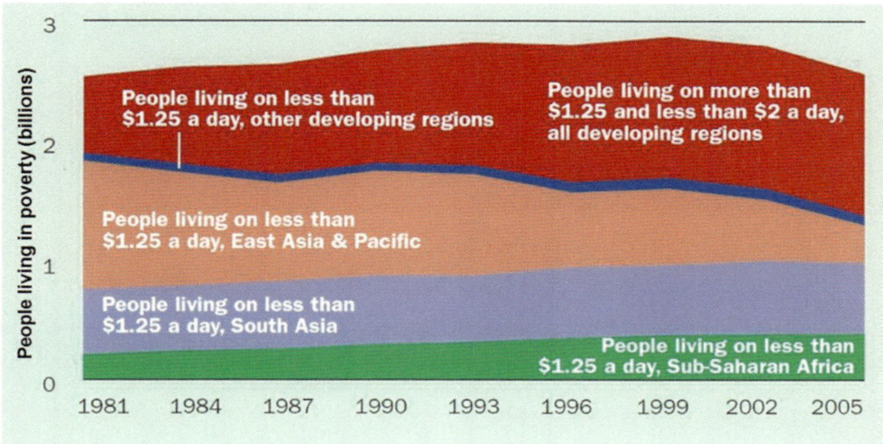

Fig. 2 Poverty in the main world regions, 1981–2005 (Source: World Bank 2008)

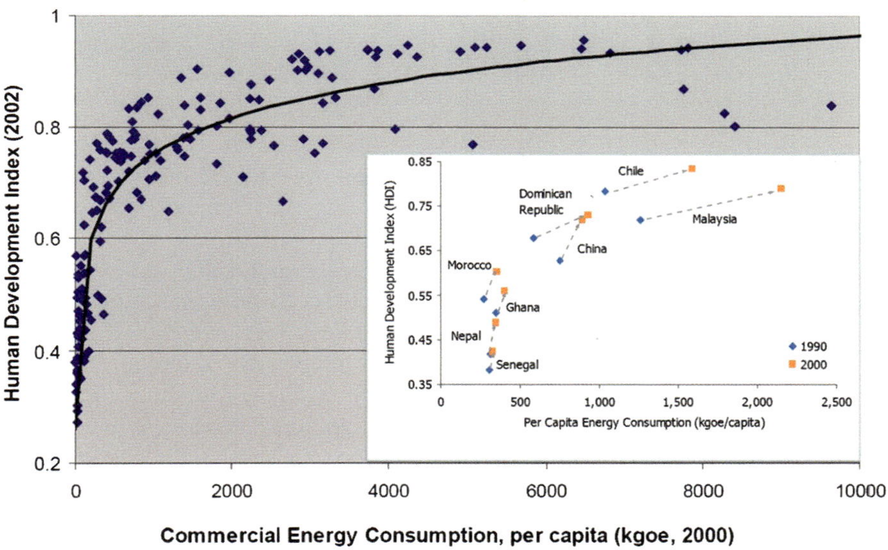

Fig. 3 The relationship between energy consumption and the Human Development Index (Source: Modi 2004; Agbemabiese 2008; Note: *kgoe* kilograms of oil equivalent)

The relationship between the Human Development Index (HDI) and commercial energy consumption per capita is illustrated in Fig. 3. There is a direct relationship that associates low energy consumption with low HDI, and relatively high energy consumption with high HDI. Analysts have found that there are low-elasticity regions where energy consumption increases do make a difference, but not as much

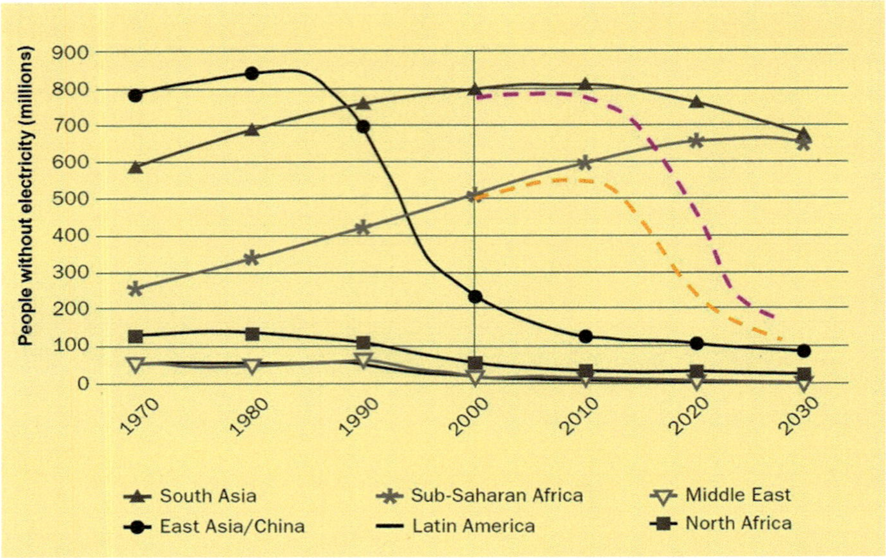

Fig. 4 Access to electricity in main world regions between 1970 and 2030 and thought experiment done in the preparatory stages of GEA (Sources: Modi et al. 2005; based on IEA 2002; GEA forthcoming; Note: *the dashed lines* indicate the fast-track approach 'Electricity for all by 2030')

as in the high-elasticity regions, where small improvements in the availability of modern energy can make a big difference to the human development indicators and the living conditions of large numbers of people.

The inset in Fig. 3 illustrates the relationship between the HDI and energy consumption in eight developing countries, and how these countries have gone through transitions over time. In Ghana, for example, between 1990 and 2000 there was a relatively small increase in per capita energy consumption, but a significant increase in our HDI. Similar patterns can be seen in China, Chile, Dominican Republic, Malaysia, Nepal and Senegal, where even more significant increases in energy consumption have been correlated with rising HDIs, demonstrating the importance of energy for reducing poverty. The MDGs forced energy analysts to go back to the drawing board and to conduct these studies.

Recent studies present projections of how access to modern energy services will change over the next few decades, up to 2030 and beyond. A study by Modi et al. (2005) for the Millennium Project shows the dramatic case of East Asia and China, where the number of people without access to electricity drops significantly, from over 800 million to just above 200 million in two decades, from 1980 to 2000 (see Fig. 4).

China has shown that with innovative institutional approaches, a wide range of new technologies, and strong government commitment and leadership, the number of people without access to modern forms of energy such as electricity can be reduced significantly – to levels low enough to make living conditions much more favourable. But in sub-Saharan Africa, the numbers of people who lack access to energy are projected to increase all the way to 2030. The same trend is evident in South Asia, with a small dip projected to occur between 2010 and 2020. Considering the statements about the importance of energy for development, it is very worrisome that according to these projections, the number of people in sub-Saharan Africa and South Asia without access to electricity will increase.

3 Securing Access to Energy

The GEA is an analysis coordinated by the International Institute for Applied Systems Analysis (IIASA), involving major energy organisations and experts from around the world (GEA forthcoming). This study examined various scenarios for reducing the numbers of people lacking access to modern forms of energy. Part of this exercise involved a thought experiment to look at a fast-track approach to achieve 'electricity for all by 2030'. Figure 4 illustrates several relevant scenarios. This thought experiment suggests that sub-Saharan Africa and South Asia should be able to replicate what China has achieved over the last few decades. This is a tall order because anything on this scale will require a global effort. It will require large commitments both from the developing countries in Africa and South Asia, but also from the developed countries and the newly emerging economies like Brazil and China. A lot of new thinking will be required, as well as the use of new policy instruments and the creation of new institutions. Analysts will have to learn quickly from experience and make the new knowledge readily available to governments and other actors in the poorer developing regions.

Equally, if not more important, considerable investments will be required. A study conducted for the Economic Community of West African States (ECOWAS) concluded that something in the order of US$50 billion was required over a 10-year period (ECOWAS 2006). That study covered only the ECOWAS region, which accounts for about a quarter of Africa. This means that several hundred billion dollars will be required over the next 10 years to make a significant dent in the number of Africans without access to modern forms of energy.

Clearly, the energy community needs a multi-track approach in every sense of the word, one that brings together the realms of technology, policy and financing. We need to draw on all the resources that are available to the global community. Figure 5 shows just the technological side, looking at greatly increasing access to electricity. Solar energy projects in remote villages are a good start, and there is a long list of reasons why decentralised solar electricity is being provided for poor villages. Yet there is no question that grid electricity is the universally preferred option, and it will therefore have to feature strongly in any electricity access programme.

My own country, Ghana, has made significant progress over the last 15 years, increasing electricity access from about 20% to over 50% of the population. This has been achieved primarily with grid electricity. But we are now approaching the

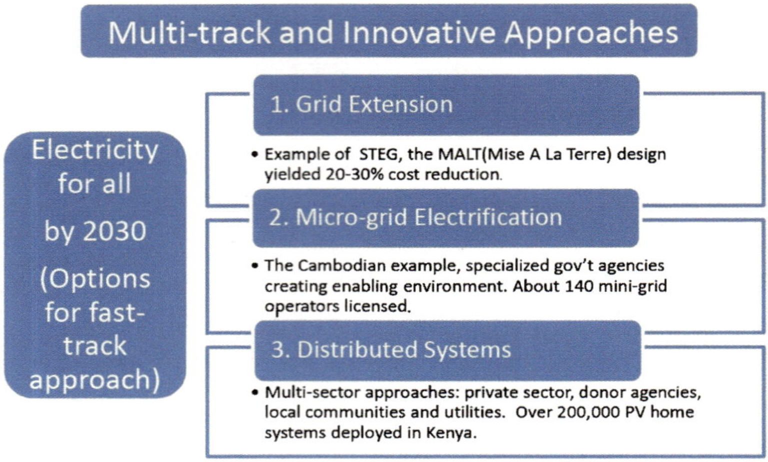

Fig. 5 Strategies to improve access to electricity (Note: *STEG* solar thermal electricity generation, *PV* photovoltaic)

limits of grid electricity, and we have to look more seriously at decentralised options in order to reach communities that are far from urban areas, far from the transmission systems, and even inaccessible in some seasons of the year. Micro-grid electrification is now being considered seriously. Distributed systems that employ known technologies, from renewables to more conventional ones, are being considered in the next phase as Ghana goes for full electrification by 2020.

Figure 6 presents the projections of the International Energy Agency (IEA 2006) showing the dependence on traditional forms of energy (solid lines). India is projected to see some increase in the number of people relying on traditional biomass, and in Africa, the increase is even larger. China has dramatically reduced the number of people without access to electricity but, as Fig. 6 shows, projections for their reduction in the use of traditional biomass are less impressive.

The thought experiment carried out for the GEA suggests that significantly reducing the dependence on traditional biomass might be feasible (dashed lines in Fig. 6). However, in order to achieve full access to modern energy systems for cooking and heating, it will be necessary to do something that even China has been unable to do over the last several decades. Drastic reductions in the use of traditional biomass are needed in India and sub-Saharan Africa, as well as in China and other parts of South Asia.

There has to be a multi-pronged approach at all levels, as shown in Fig. 7. Evidence across the board shows that without strong leadership from national or regional governments one will not make much of a difference. The multi-pronged approach required will therefore have to involve pressing national governments to take action. This is essentially a demand issue: one can make all the resources available, but if national and local governments are more interested in misappropriating the money, one will not get the needed results. We have to identify ways to stimulate

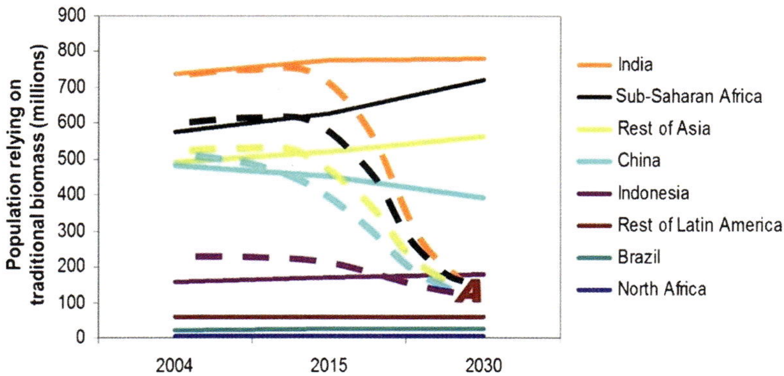

Fig. 6 Reducing the dependence on traditional biomass between 2004 and 2008 and thought experiment in the preparatory stages of GEA (Sources: Based on IEA 2006; GEA forthcoming; Note: *the dashed lines* indicate the fast-track approach 'Cleaner cooking and heating systems for all by 2030')

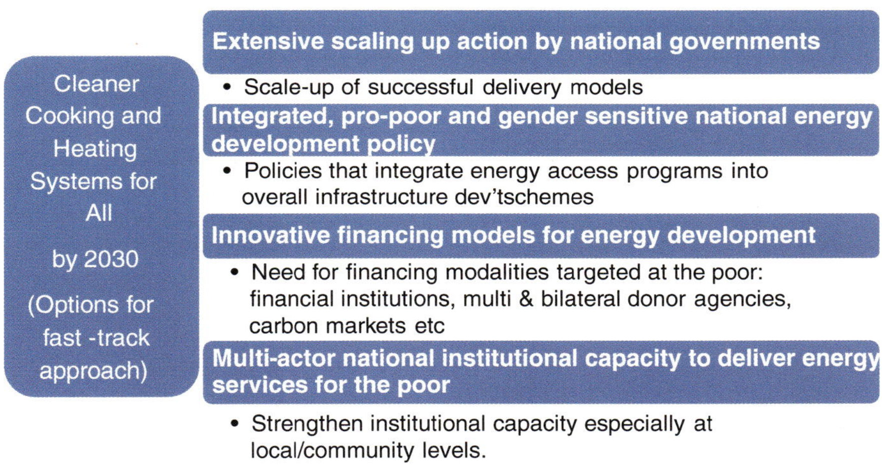

Fig. 7 Strategies to improve access to cleaner cooking and heating systems

national governments to champion the cause of making cleaner energy available to their peoples for cooking and heating, including productive uses of energy.

As Schelling (2012) convincingly points out, the current energy situation cannot continue. The global community will have to commit itself to eradicating energy poverty in the coming decades. This means that all the best brains and muscle of this world will need to be brought on board, and substantial financial resources – 1% of global GDP or more – will need to be mobilised. The big challenge facing the world today is to garner the courage to commit to such a bold initiative and to mobilise the financial resources, knowledge and institutional resources needed to achieve this ambitious goal.

At a forum of African energy ministers a couple of years ago, I presented statistics on access to modern forms of energy, in this case for electricity, for Africa, and noted that there were some 20 countries where access levels are still less than 10%. It is simply unacceptable that in any country today only 3 or 4% of the population have access to electricity. We need to use all options in the book to increase the pressure on governments to act.

We also need to make use of innovative financing models. We need to draw in the funding, including local financing, and to catalyse funding from the industrialised world and newly emerging economies. But we also need to use a wide range of schemes, including smart subsidies and time-bound mechanisms, to ensure that we reach the poorest of the poor. Next, there is the big question of the institutional capacity and the institutional forms that will be required to carry this effort forward. This goes all the way from the global to the subregional and community levels, since local institutions and a wide range of local actors will be needed to implement these schemes. New initiatives are needed to draw in businesses of all sizes, ranging from the large energy companies to small individual micro-enterprises. Many of us who have lived in Africa will recall that even in urban areas it was possible to buy kerosene from street vendors who sold it in bottles carried on top of their heads. We need to draw on the whole range of institutional options to be able to deliver the access levels that are required.

4 Charting the Way Forward

What would be the most desirable features of energy access programmes? It is important that we do not forget the issues that have been raised in the recent past. The number one goal of any energy access programme must be to ensure that energy is available for productive uses. The MDGs aim to eradicate extreme poverty; they are about enabling people to generate incomes that will improve their wellbeing so that they can afford social services, including energy services.

The next key issue is to scale up energy supply capacity. To increase access to electricity, for instance, one needs to increase the power supply and to draw on the full range of generation options, all the way from large-scale systems to small decentralised ones, including small hydro and solar home systems. It is important to maintain the widest range of technological options in any future programme. On the financing side, smart subsidies that are time-bound and well-targeted will be required. This has been a big challenge for many developing countries. Efficient pricing and public–private partnerships will also have to be major features of any future energy access programme.

What I am calling for is something that has emerged from the experience with achieving the MDGs to date. That is the sobering realisation that energy is at the very heart of achieving the MDGs. I am calling for a commitment from the global community to launch a comprehensive programme for achieving universal energy access by 2030. Ambitious as it is, achieving universal access would make the difference in ensuring global security, and in improving the livelihoods of one-third of the world's population.

Achieving 100% access to electricity will have to be part of this global energy access agenda. Providing 100% access to cleaner cooking and heating systems will also have to be part of it, emphasising the full range of options. This does not mean that everyone will be using liquefied petroleum gas (LPG) by 2030 – that would be nearly impossible. But many people will be using improved cooking stoves that are more efficient and that feed into local economic systems because they would grow their own wood and increase their incomes.

It is important to see these targets in the broad context. Emphasis on sustainable energy for productive uses must be the cornerstone of the future commitment to a global agenda for universal energy access by 2030.

References

Agbemabiese, L. (2008). Renewable energy enterprise development. In A. Brew-Hammond & F. Kemausuor (Eds.), *Renewable energy for rural areas in Africa – The enterprise development approach* (pp. 6–14). Kumasi: Kwame Nkrumah University of Science and Technology (KNUST) University Press.

Bazilian, M., Sagar, A., Detchon, R., & Yumkella, K. (2010). More heat and light. *Energy Policy, 38*(10), 5409–5412.

ECOWAS (Economic Community of West African States). (2006). *White paper for a regional policy*. Abuja: ECOWAS. www.gm.undp.org/Reports/ECOWAS%20energy%20white%20paper.pdf. Accessed 3 May 2010.

GEA (Global Energy Assessment). (forthcoming). *The global energy assessment: Toward a more sustainable future*. Laxenburg/Cambridge: IIASA/Cambridge University Press.

Halsnæs, K., & Garg, A. (2011). Assessing the role of energy in development and climate policies: A conceptual approach and key indicators. *World Development, 39*(6), 987–1001.

IEA (International Energy Agency). (2002). *World energy outlook 2002*. Paris: OECD/IEA.

IEA (International Energy Agency). (2006). *World energy outlook 2006*. Paris: OECD/IEA.

Irvine-Halliday, D., Doluweera, G., Platonova, I., & Irvine-Halliday, J. (2008). SSL (solid state lighting): A big step of the poverty trap for the bop (bottom of pyramid)! *Journal of Light and Visual Environment, 32*(2), 258–216.

Modi, V. (2004). *Energy services for the poor.* Paper commissioned by the Millennium Project Task Force 1. www.unmillenniumproject.org/documents/MDPpaperdec31.pdf. Accessed 29 Apr 2010.

Modi, V., McDade, S., Lallement, D., & Saghir, J. (2005). *Energy services for the Millennium Development Goals* (UN Millennium Project). New York: UNDP, World Bank and ESMAP.

Schelling, T. (2012). Rethinking energy aid mechanisms: Three premises. In F.L. Toth (Ed.), *Energy for development: Resources, technologies, environment* (pp. 27–34). Dordrecht: Springer.

UN (United Nations). (2000). *United Nations Millennium Declaration* (Resolution 2, session 55, 8 September 2000). New York: UN General Assembly.

WHO & UNDP (World Health Organization & United Nations Development Programme). (2009). *The energy access situation in developing countries: A review focusing on the least developed countries and Sub-Saharan Africa*. New York: WHO/UNDP.

World Bank. (2008). *Poverty data: A supplement to world development indicators 2008*. Washington, DC: World Bank.

Multilateralism and Energy for Development

Kandeh K. Yumkella

Abstract The lack of access to affordable and reliable energy services is a key obstacle to human, social and economic development. Yet billions of people worldwide lack access to electricity and still rely on traditional biomass as their primary source of energy. This chapter presents a vision for scaling up efforts to provide the poor with modern, safe and efficient energy services in the context of energy–development linkages, climate change and the geopolitics of energy security. Energy for development strategies should aim at transformative changes that bring about sustainable development and address climate change in a holistic manner because a transition to low-carbon economies is a pressing imperative. Energy security and reliable access to relatively inexpensive energy supplies is a prerequisite for economic development and competitiveness. New institutional arrangements are needed to address these challenges.

1 Context

In the course of its work, the United Nations Industrial Development Organization (UNIDO) is constantly reminded by Member Countries of the paramount importance of energy for development. At the global level, large parts of the population – billions of people – have no access to modern energy services. These are services that most people in industrialised countries take for granted, such as adequate lighting, low-polluting energy for heating and cooking, and motive power. About 1.4

K.K. Yumkella (✉)
UN Secretary General's Advisory Group on Energy and Climate Change, UN-Energy,
United Nations Industrial Development Organization (UNIDO),
Wagramerstr. 5, P.O. Box 300, A-1400, Vienna, Austria
e-mail: k.yumkella@unido.org

F.L. Toth (ed.), *Energy for Development: Resources, Technologies, Environment*,
Environment & Policy 54, DOI 10.1007/978-94-007-4162-1_4,

billion people still do not benefit from access to electricity, and around 2.7 billion rely on the use of so-called traditional biomass as their primary source of energy (IEA et al. 2010). This represents an unacceptable and unsustainable reality. It is widely accepted that the lack of access to affordable and reliable energy services is a key obstacle to human, social and economic development. Modern energy represents a critical foundation for sustainable development and for achieving the Millennium Development Goals (MDGs) (UN-Energy 2005).

Despite the significant efforts by local institutions and governments, utilities and international organisations, the absolute number of energy poor is expected to rise in coming decades. History has shown, however, that significant progress can be achieved with regard to improving energy access within a short timeframe. Remarkable improvements have occurred in several Asian countries, South Africa and Brazil in the recent past. However, current initiatives to eradicate energy poverty are insufficient in scale and scope, and attempting to address the issue in the same manner as in the past is clearly inappropriate.

The issue of energy access is receiving greater and greater attention. It is crucial to capitalise on this momentum, as energy is central to facing many of today's key development challenges. Addressing the issue of energy poverty in a comprehensive manner would have enormous multiple benefits (Bazilian et al. 2010a). This chapter presents a vision, forged through many years of dedicated service, for moving this agenda forward and scaling up efforts to provide the poor with modern, safe and efficient energy services. I start by considering the energy–development nexus in greater detail. The third section considers energy in the context of climate change. Geopolitics and energy security are discussed in the fourth section. The final section examines multilateralism in the energy field and its evolution.

2 The Energy–Development Nexus in an Evolving World

Lack of energy access and poverty are closely correlated (see Fig. 1). It is no coincidence that the poorest people in the world also happen to be those without access to adequate energy services. Access to affordable energy services is a prerequisite for socio-economic development. Energy is central to the development and poverty reduction discourse. Dr Rajendra Pachauri, Chair of the Intergovernmental Panel on Climate Change, put it most succinctly: '*Energy remains the missing MDG*' (Reuters 2009).

There has been significant progress in alleviating energy poverty in certain developing regions, but such improvements are far from what is needed. Indeed, without more effective actions, the absolute number of energy poor is expected to increase in certain regions, most notably in sub-Saharan Africa (IEA 2009). Changing this trend requires an international political commitment that sets out targets, actions and associated benchmarks (Bazilian et al. 2010b).

The recent succession of economic crises took a harsh toll on the world's poor. First, the food price crisis pushed millions of people back into a state of hunger

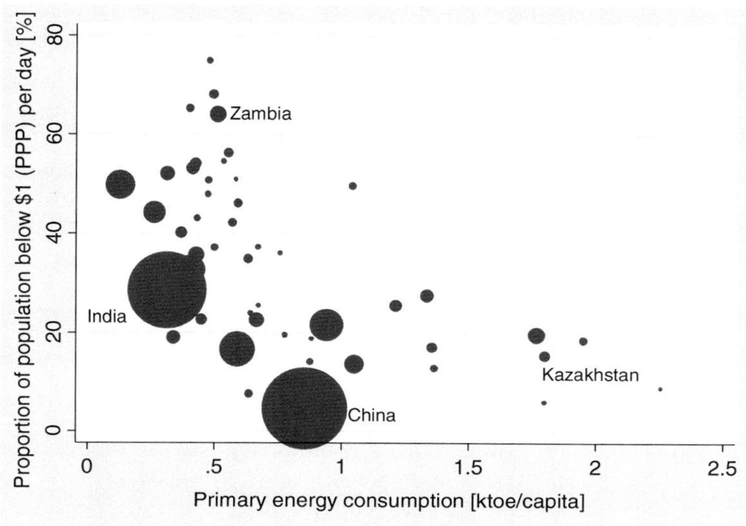

Fig. 1 The energy–poverty nexus (Sources: UNSD et al. n.d.; UNSD n.d.; IEA 2007; Note: The size of each dot is proportional to the country's population. *PPP* purchasing power parity, *ktoe* kilotonnes of oil equivalent)

(FAO 2009), reversing in a few months the hard-won progress gained over decades. Volatile energy prices also adversely affected the poor disproportionately. In addition, the financial crisis drastically reduced the availability of much-needed capital to invest in ageing energy systems already challenged by increased energy demand and population growth.

This rapidly changing landscape has challenged the development community and has required swift adjustments in the orientation of development policies. There was a powerful convergence of these adverse effects on the poor across the developing world. These changes may wipe out some of the MDGs' greatest accomplishments. Indeed, the World Bank estimates that the present economic crisis has pushed some 53 million more people into poverty in developing countries (World Bank 2009).

I believe that socio-economic development requires wealth generation. There is no empirical evidence of poverty reduction without an increase in economic activity. Once basic energy services (e.g. for cooking and heating) are available, there is a strong case for focusing the provision of additional energy on income-generating activities. By creating wealth, a virtuous cycle is initiated. Not only can the poor afford the energy that ensures the financial sustainability of the service, they also gain the means to improve their livelihoods. Poverty alleviation requires sustained economic growth, which is fuelled in part by new energy supplies.

This is where energy for productive uses comes into play. Many organisations are raising money to help African communities install single solar panels to provide lighting in their homes. But, in many cases, this is simply shining a light on their poverty. Energy should be provided so that the poor can enhance their productivity and develop their economies. In the short-term, engineers need to supply simpler technologies to reach this goal. Energy sources are needed so that women can process and prepare food more efficiently, saving time and physical energy. Communities should have sufficient energy to power their clinics, small businesses and their schools – not just their homes.

Energy for productive uses is critical because of demographic trends and the need to feed an increasing population. The United Nations estimates that by 2030, sub-Saharan Africa's population may be between 1.2 and 1.5 billion people, roughly the size of China's current population, and that by 2050 it could grow to two billion or more (UN 2009). Over half of the population will live in cities that are already overcrowded, such as Freetown in Sierra Leone and Lagos in Nigeria. Urbanisation, combined with expected changes in diets, will require significant increases in national and worldwide food production.

Without energy for productive uses in these poor countries, what will happen to all of these urban dwellers? Humans are ingenious, and they are mobile. There is a likelihood that our African grandchildren will head for Europe and North America. The world does not need these African people settling as undocumented aliens. It needs them to come as productive and educated citizens. It is in the collective interest to prevent this demographic crisis by developing energy systems that help to generate new income so people remain where they want to be.

A significant share of the world's population cannot continue to rely primarily on incomes from agriculture. Development strategies must help to transform their economies. Yet, greater energy requirements must be met while addressing climate change.

3 Energy and Climate Change

Energy-related activities account for two-thirds of all greenhouse gas emissions (see Fig. 2). Here again, energy is central to a major global issue, climate change, arguably the biggest threat humanity is facing. A broad-ranging energy transition is required to mitigate climate change. To address the challenge at hand will require not an incremental evolution but an all-encompassing paradigm shift that will lead towards low-carbon economies.

Poor people care about climate change because it is already affecting them, whether they live in India, Sierra Leone, Pakistan or the Philippines. In those countries, new technologies as well as technical and financial assistance are needed. The international community must not fail to meet their aspirations. The asymmetry in climate change impacts and the historical responsibility for greenhouse gas emissions have resulted in expectations for *carbon justice*, as advocated by the late Wangari Maathai (2010), Nobel Laureate in 2004 and environmentalist from Kenya.

Inclusive globalisation that reduces poverty and ensures the spread of prosperity must put energy at the centre of deliberations and actions. Several international institutions are leading the way.

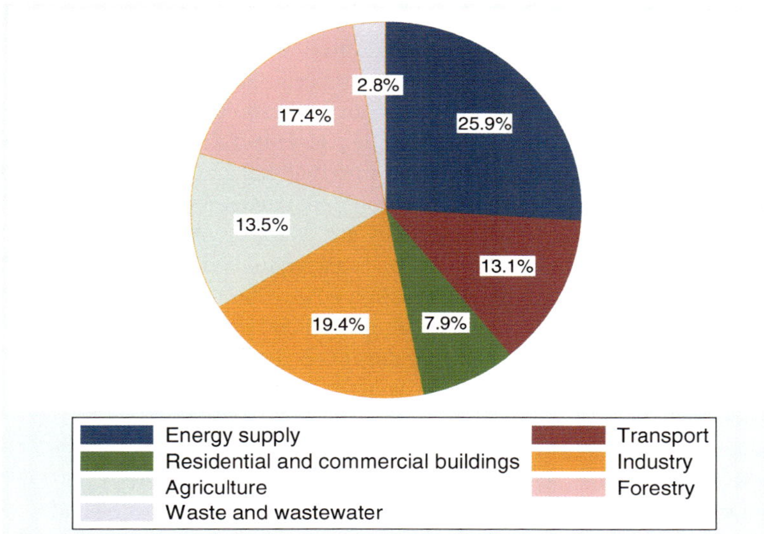

Fig. 2 Greenhouse gas emissions by sector, 2004 (Source: Barker et al. 2007)

Climate change and energy access are intrinsically related. Energy is essential for reducing poverty. But at the same time, actions to drastically reduce the anthropogenic interference with the climate system need to be taken. Providing the poor with climate-resilient systems based on local, sustainable sources is no utopia. Improved commercially available technologies are needed to leapfrog expensive and polluting systems and allow clean energy services for the poor at low cost.

4 Geopolitics and Energy Security

Representatives of developing countries at international forums (ministers, ambassadors, or other government officials) often emphasise that the geopolitics of energy should not be underestimated. Such considerations are at the core of the linkages between energy, security, climate change and poverty.

The Gulf of Guinea, the region extending from Angola to Mauritania, is expected to supply 15–20% of the oil and gas of some developed countries. This region is, however, one of the most conflict-ridden in the world. Environmental degradation is severe and mounting. Although Nigeria is the fifth largest supplier of oil to the United States, power blackouts are frequent. A holistic approach to assist these regions must encompass political issues but also resources and pollution management, as well as the promotion of economic opportunities. Energy is central to local and global geopolitical stability.

Many developing countries are considering civilian nuclear power as an option to address some of their energy needs. Several of them have significant uranium resources to exploit. The issue of resource extraction, including oil and gas, is central to political discussions about transforming these economies into competitive and viable players at regional and global levels. But if oil and gas are handled as mere commodities like cocoa or coffee, a major global crisis is much more likely.

Energy security is not merely of concern to industrialised countries. The economic implications of unreliable energy services are daunting. A number of developing countries suffer from inadequate generation capacity, limited electrification, inefficient systems and high production and distribution costs. Chronic energy problems take a heavy toll in terms of losses in economic growth and productivity. Estimates of the cost of power shortages reveal that they may exceed several per cent of GDP (Foster and Briceno-Garmendia 2010). It is critical to include qualitative aspects of energy services in the consideration of energy poverty alleviation.

Trade is another sensitive dimension of energy that requires serious attention. Trade in energy carriers, technology transfer, intellectual property rights and institutional frameworks all have widespread implications. They must be addressed in a comprehensive fashion. Developing countries must not follow the dirty development path taken many decades ago by the now industrialised countries.

5 Multilateralism on Energy and its Evolution

The thinking in the development community has evolved remarkably over recent decades. In the mid-twentieth century, the focus on the accumulation of physical capital as the primary means to catch up with industrialised countries gradually gave way to the recognition of the importance to invest in human capital. This trend was followed by ideas derived from the Washington Consensus that pointed towards accelerating structural changes, liberalisation, deregulation and privatisation. However, the overwhelming emphasis on macroeconomic orthodoxy did not always deliver on its promise to promote sustainable and equitable development. This realisation led to a series of reforms addressing institutional shortcomings.

The lesson that can be drawn from the past is that development strategies should encompass, *inter alia*, well-designed, context-specific policies, macroeconomic stability and a flexible institutional framework. The contemporary development agenda has become broader and more complex. There is a common understanding in the development community that a number of major challenges are intrinsically interrelated.

The concept of first and third worlds, with its implicit reference to donors and recipients in the context of development cooperation, belongs to history and is inappropriate now. We live in a multi-polar, rapidly evolving world. Multilateralism must therefore evolve to better serve the mutual interests and responsibilities of both industrialised and developing countries, while recognising different perspectives and circumstances. Modern multilateralism is about engaging with new players, harnessing global and regional institutions to help address threats and seizing opportunities that surpass the capacities of individual states (Zoellick 2010).

This evolving context has tremendous consequences for the way the international community addresses energy-related challenges. In that regard, and in line with the proposal of Mohamed ElBaradei, former Director General of the International Atomic Energy Agency (IAEA), to create a global energy organisation (ElBaradei 2008), energy issues need to be addressed in a coordinated and holistic fashion at the institutional level.

Various international organisations are involved in the energy sector. Yet, most of them are specialised, either in terms of energy sources (e.g. IAEA; Organization of the Petroleum Exporting Countries, OPEC; International Renewable Energy Agency, IRENA), narrow constituencies (e.g. International Energy Agency, IEA), or focus on particular aspects (e.g. UNIDO; Food and Agriculture Organization of the United Nations, FAO). This institutional landscape has been complemented by a range of other bodies (e.g. Major Economies Forum on Energy and Climate, MEF) with various mandates.

The UN system has worked on energy issues for decades, but in a rather fragmented manner. Responding to the need for a concerted UN strategy on energy issues, UN-Energy (see Box 1) was established. This inter-agency mechanism for coherence and cooperation on energy issues aims at effectively combining the experience of some 20 UN agencies to implement sustainable energy solutions.

Box 1 UN-Energy

At the World Summit on Sustainable Development (WSSD), held in Johannesburg in 2002, it became clear to delegates that poverty reduction, access to energy, energy security and climate change mitigation were all interlinked issues requiring a coordinated response from the development community. UN-Energy was established in 2004 to help ensure coherence in the United Nations system's multidisciplinary response to the WSSD, and to support countries in their transition to sustainable energy.

The core fields of access to energy, renewable energy and energy efficiency – UN-Energy's clusters – have garnered major attention and experienced rapid growth in investments and policy-related focus with an ever-growing number and variety of players involved. UN-Energy aims to promote system-wide collaboration in the area of energy with a coherent and consistent approach, as there is no single entity in the UN system that has primary responsibility for energy. Its role is to increase the sharing of information, encourage and facilitate joint programming and develop action-oriented approaches. It was also initiated to encourage collective engagement between the United Nations and other key external stakeholders, including the private sector. UN-Energy brings together members on the basis of their shared responsibility, deep commitment and stake in achieving sustainable development.

Since its inception in 2004, UN-Energy has been instrumental in sharing information and good practices on energy within the UN system. It has also served as a platform for partnerships among various UN organisations and for coordinating their energy-related programmes and activities. Since there is no single entity that has primary responsibility for energy, UN-Energy is vital for a focused and collective UN engagement across the energy field.

Several years have passed since UN-Energy's establishment in 2004. During these years energy has dramatically moved to centre stage on the global agenda. This is reflected in its increased prominence in global security, environmental and development debates. The clear understanding that energy is an essential part of addressing all of these issues is further underlined by the fourfold increase in annual global clean energy investments – from US$36 billion when UN-Energy was established to US$145 billion in 2009 (WEF 2009).

Concerted global efforts are required to develop and implement smart policies. Fortunately, a broad range of options that can effectively address energy-related challenges exists today. As an example, modern systems based on renewable energy can be an effective tool for helping the 2–3 billion people worldwide who still lack access to energy services. Likewise, energy efficiency measures are now well understood and can serve as a foundation for national energy plans.

The United Nations has responded to the challenges and opportunities in the energy sector with numerous programmes and projects addressing key areas from financing to capacity development and technical assistance (UN-Energy 2010a). An external review of UN-Energy was recently completed. The report of the review presents a number of options for enhancing the delivery capacity of UN-Energy to better serve the global community and provide a strong foundation for collective and coordinated action (UN-Energy 2010b).

The task of coordinating over 20 international organisations with different agendas is daunting. Yet, I believe in UN-Energy's mandate and capacity to deliver, as I also believe that the need for a clear, coherent and coordinated engagement of the UN system in energy is even more compelling nowadays than ever before. Utilising the different perspectives and comparative advantages of the various agencies in a coherent fashion is likely to produce the richest and most comprehensive set of actions to match the complexity of the challenges faced.

As one concrete example, the IAEA develops and disseminates some of the best energy planning tools (IAEA 2009). Significant efforts have been put into customising generic modelling frameworks to specific needs, such as those of poor developing countries. A number of capacity building projects are ongoing, with the objective of training national experts in energy planning, an essential element of energy policy design and decision making. Two demonstration case studies analyse alternative policies to increase the share of renewables in the energy supply mix in Ghana (UN-Energy 2006) and Sichuan, China (UN-Energy 2007). The reports present analytical results from inter-agency cooperation (IAEA; UN Department of Economic and Social Affairs, DESA; FAO; United Nations Environment Programme, UNEP; and UNIDO). Coordinating and partnering with other UN-Energy Members has helped refine the tools and reach out to a broader

Box 2 UN Secretary-General's Advisory Group on Energy and Climate Change (AGECC)

In June 2009, the UN Secretary-General Ban Ki-moon launched the AGECC. Given the increasing political priority to energy and climate change, the promotion of climate-friendly energy solutions and adequate and reliable supplies of energy should be addressed in tandem in a broader context of national development priorities and the Millennium Development Goals. The UN is leading a process to ensure high-level political support in the climate change negotiations.

To this end, the Secretary-General must have at hand a comprehensive and up-to-date assessment not only of the challenges that developed and developing countries face in effectively addressing climate change, but also the potential role that the UN as a whole can play in helping developing countries meet these challenges. The group is expected to consider ways to comprehensively address climate change, energy poverty and boost energy efficiency and clean production, with a focus on developing countries. The AGECC includes not only prominent UN figures, but also policy, technical and business leaders from the private sector and research institutions from both industrialised and developing countries.

audience, thereby increasing the impact in terms of capacity building in a number of countries.

In addition to this, Ban Ki-moon, Secretary-General of the United Nations, decided to launch another body in June 2009, the Advisory Group on Energy and Climate Change (AGECC) (see Box 2). While the UN system is well placed to play a major role, its achievements can be greatly increased by joining hands with other actors.

The AGECC aims at uniting cutting-edge expertise and experience related to sustainable energy. Its members include the heads of some of the world's most successful energy companies. The visionary thinking of these world leaders can help to chart a clear course for the UN with regard to energy and climate change issues.

The AGECC is a prime example of a multi-stakeholder partnership that brings together the UN system, including the World Bank, with the private sector and research institutions. Its work has benefited from a unique mix of policy orientation, technical expertise and business experience of leading figures in the field of energy. After a number of meetings, the AGECC produced a report that was presented to the Secretary-General. The report calls on the United Nations and its Member States to commit themselves to two complementary goals (AGECC 2010), namely, to ensure universal access to modern energy services by 2030; and reduce global energy intensity by 40% by 2030.

The report makes a couple of points clear. First, it is unacceptable that a third of humanity has no access to modern energy services and that half of humanity has to rely on traditional biomass to meet their basic energy needs. Eliminating energy poverty is of paramount importance in eradicating poverty. It is also essential to the achievement of the other Millennium Development Goals. Second, the vast potential for energy efficiency improvements across the energy supply and delivery chain remains largely untapped.

The AGECC has therefore called for commitment and concerted action on the two ambitious but achievable goals described above. To this end, the group recommends actions towards achieving these goals:

- A global campaign should be launched in support of 'Energy for Sustainable Development'.
- All countries should prioritise these goals through the adoption of appropriate national strategies.
- Finance, including innovative financial mechanisms and climate finance, should be made available by the international community.
- Private-sector participation in achieving the goals should be emphasised and encouraged.
- The UN system should make 'Energy for Sustainable Development' a major institutional priority.

The UN system can catalyse these actions by establishing a mechanism to track progress towards the goals and by providing the requisite support to strengthen national capacities to achieve them. Institutionally 'embedding' the energy-related goals in the work of the UN system would help sustain efforts towards the achievement of the goals in the long term. UN-Energy is well positioned to be the hub for such collective engagement.

The Secretary-General remains committed to the issues of universal access to modern energy services and energy efficiency. Collective endeavour under the leadership of the Secretary-General offers enormous opportunities for positive change.

6 Conclusions

Energy for development strategies must go well beyond merely providing light to poor households. They should aim at transformative changes that bring about sustainable development. The recent succession of crises has set back some development progress. International organisations and the UN system need to adjust swiftly to the new circumstances and provide advice and assistance that is resilient and long-lasting, and creates an environment that is conducive to enhancing endogenous development. Today, there is no technical barrier to providing the billions of energy poor with modern, safe, reliable and affordable energy services. It is our duty to deal with the aspiration of countries to move towards modern economies, and energy is paramount to such transformation.

There is no solution to climate change without addressing the energy challenges in a holistic manner. Nothing short of a paradigm shift is required, commonly referred to as an *energy revolution* by the International Energy Agency (IEA 2009). In other words, we need to profoundly revisit the ways we produce, transport and consume energy. The stakes are high, and poor developing countries are already paying the price of problems they barely contributed to. A transition to low-carbon economies is a pressing imperative.

Energy security is at the centre of many countries' concerns today. Reliable access to relatively inexpensive energy supplies is a prerequisite for economic development and competitiveness. It is therefore no surprise that in industrialised and developing countries alike, energy security plays a predominant role in decision-making and policy processes.

The institutional arrangements to address those challenges are crucially important. Equally relevant is the ability of institutions to adapt and deal with the evolving context. The UN and its partners are calling for the scaling up of global efforts in clean, efficient energy that brings affordable energy services for all. I fully agree with the Secretary-General that doing so would take us a long way towards mitigating climate change and achieving the MDGs.

Acknowledgments My gratitude goes to the organisers of the IAEA Scientific Forum entitled 'Energy for Development' at the IAEA General Conference 2009 for bringing such a crucial topic to the attention of Member States and other stakeholders. I am thankful to Patrick Nussbaumer and Morgan Bazilian, my special advisor on energy issues, from UNIDO for assisting me in consolidating my presentation into this chapter.

References

AGECC (Advisory Group on Energy and Climate Change). (2010). *Energy for a sustainable future*. New York: The UN Secretary-General's Advisory Group on Energy and Climate Change. www.unido.org/fileadmin/user_media/Services/Energy_and_Climate_Change/EPP/Publications/AGECC_Report.pdf

Barker, T., Bashmakov, I., Bernstein, L., Bogner, J. E., Bosch, P. et al. (2007). Technical Summary. In B. Metz, O. R. Davidson, P. R. Bosch, R. Dave, & L. A. Meyer (Eds.), *Climate change 2007: Mitigation* (Contribution of Working Group III to the fourth assessment report of the Intergovernmental Panel on Climate Change). Cambridge: Cambridge University Press. www.ipcc.ch/pdf/assessment-report/ar4/wg3/ar4-wg3-ts.pdf

Bazilian, M., Sagar, A., Detchon, R., & Yumkella, K. (2010a). More heat and light. *Energy Policy, 38*, 5409–5412. doi:10.1016/j.enpol.2010.06.007.

Bazilian, M., Nussbaumer, P., Cabraal, A., Centurelli, R., Detchon, R. et al. (2010b). *Measuring energy access: Supporting a global target*. New York: The Earth Institute, Columbia University. http://modi.mech.columbia.edu/files/measuring_energy_poverty_merge_8_A_3.pdf

ElBaradei, M. (2008, July 24). A global agency is needed for the energy crisis. *Financial Times*.

FAO (Food and Agriculture Organization). (2009). *Food security and the financial crisis*. Rome: World Summit on Food Security (WSFS) Secretariat, FAO. ftp://ftp.fao.org/docrep/fao/meeting/018/k6360e.pdf

Foster, V., & Briceno-Garmendia, C. (Eds.). (2010). *Africa's infrastructure: A time for transformation*. Washington, DC: Agence Française de Développement and World Bank. www.infrastructureafrica.org/aicd/library/doc/552/africa%E2%80%99s-infrastructure-time-transformation

IAEA (International Atomic Energy Agency). (2009). *IAEA tools and methodologies for energy system planning and nuclear energy system assessments*. Vienna: IAEA. www.iaea.org/OurWork/ST/NE/Pess/assets/09-16631_iaea_tools_brochure.pdf

IEA (International Energy Agency). (2007). *World energy statistics and balances* (Database). Paris: IEA.

IEA (International Energy Agency). (2009). *World energy outlook 2009*. Paris: IEA.

IEA, UNDP, & UNIDO (International Energy Agency, United Nations Development Programme & United Nations Industrial Development Organization). (2010). *Energy poverty: How to make modern energy access universal?* Special early excerpt of the *World energy outlook 2010* for the UN General Assembly on the Millennium Development Goals. Paris: IEA. www.worldenergyoutlook.org/docs/weo2010/weo2010_poverty.pdf

Maathai, W. (2010). Commentary: carbon justice and forestation – The African perspective. In H. J. Schellnhuber, M. Molina, N. Stern, V. Huber, & S. Kadner (Eds.), *Global sustainability: A Nobel cause*. Cambridge: Cambridge University Press.

Reuters. (2009, January 21). *Energy missing Millennium goal.* U.N. climate chief, 21. www.reuters.com/article/idUSDEL270134._CH_.2400

UN (United Nations). (2009). *World population prospects: The 2008 revision*. New York: UN Department of Economic and Social Affairs. http://esa.un.org/unpd/wpp2008/peps_documents.htm

UN-Energy. (2005). *The energy challenge for achieving the Millennium Development Goals*. New York: UN-Energy. www.un-energy.org/sites/default/files/share/une/un-enrg_paper.pdf

UN-Energy. (2006). *Assessing policy options for increasing the use of renewable energy for sustainable development: Modelling energy scenarios for Ghana*. New York: DESA, FAO, IAEA, UNEP, UNIDO. www.un-energy.org/sites/default/files/share/une/ghana.pdf

UN-Energy. (2007). *Assessing policy options for increasing the use of renewable energy for sustainable development: Modelling energy scenarios for Sichuan, China*. New York: DESA, IAEA, UNEP. www.un-energy.org/sites/default/files/share/une/sichuan_case_study.pdf

UN-Energy. (2010a). *Delivering on energy: An overview of activities by UN-Energy and its members*. New York: UN-Energy. www.un-energy.org/sites/default/files/share/une/delivering_on_energy.pdf

UN-Energy. (2010b). *Looking to the future*. New York: UN-Energy. www.un-energy.org/stories/41-un-energy-looking-to-the-future

UNSD, DESA & UN (United Nations Statistics Division, Department of Economic and Social Affairs & United Nations). (n.d.). *Millennium Development Goals indicators*. http://unstats.un.org/unsd/mdg/Default.aspx

UNSD (United Nations Statistics Division). (n.d.). *UN data*. http://data.un.org/

WEF (World Economic Forum). (2009). *Green investing: Towards a clean energy infrastructure*. Geneva: WEF. www.weforum.org/pdf/climate/Green.pdf

World Bank. (2009, February 12). *Crisis hitting poor hard in developing world.* Press Release. Washington, DC: World Bank. http://go.worldbank.org/PGNOX87VO0

Zoellick, R. (2010). Modernizing multilateralism for a multipolar world. *Making It 3*, 32–33. www.makingitmagazine.net/?p=1816

The Possible Role of the Energy Charter in Advancing Energy for Development in Africa

André Mernier

Abstract One of the key challenges facing the world today is to ensure the provision of energy services to the poor. Modern energy services are essential for improving productivity, creating enterprises, increasing employment and incomes, and providing effective public services. Yet millions of people in Africa, parts of Latin America and in Asia lack even basic energy services. The Energy Charter is a dynamic and evolving community. Membership of the Treaty can foster investments in the energy sector by giving a positive message to potential investors who can be reassured by the security offered by the Treaty's provisions on investment protection. The Treaty also provides opportunities to promote cooperative solutions appropriate to local and regional markets and demands. The Treaty is open to all states that are ready to share its basic principles, a very important one for Africa being sovereignty over energy resources.

1 Energy and Development in Africa

The world today faces a number of pressing challenges in the area of energy. Among these, none is more important than ensuring the provision of energy services to the poor. The link between energy and development is well established. Modern energy services are essential for improving productivity, creating enterprises, increasing employment and incomes, and providing effective public services such as education and healthcare. Given the fundamental importance of energy for achieving sustainable development, it will not be possible to achieve the United Nations Millennium

A. Mernier (✉)
Energy Charter Secretariat, Boulevard de la Woluwe, 56, B-1200 Brussels, Belgium
e-mail: SG@encharter.org

F.L. Toth (ed.), *Energy for Development: Resources, Technologies, Environment*, Environment & Policy 54, DOI 10.1007/978-94-007-4162-1_5,

Development Goals by 2015, as agreed by the world's leaders, without improving the provision of energy to the world's poorest societies.

Despite the imperative to tackle energy poverty, significant numbers of people in Africa, parts of Latin America and in Asia continue to lack even basic energy services. Improving access to electricity is a particular challenge for many developing countries, where about 1.6 billion people are without electricity services. The large-scale electrification schemes currently underway in many middle-income nations and in the poorer countries of Asia are likely to improve the situation in these countries. However, other parts of the world remain desperately in need of practical initiatives to increase electricity access.

In sub-Saharan Africa, especially in the rural areas, the situation is particularly acute. The World Bank estimates that only 51% of urban and 8% of rural households have an electricity connection. The average per capita consumption of electricity in Africa (excluding South Africa) is just 124 kWh/year, barely 1% of the typical consumption in high-income countries. Without access to modern and sustainable energy services, the poor in Africa are deprived of opportunities to improve their living standards.

According to all projections, less than half of African countries will have achieved universal access to electricity by 2050. With current trends indicating that Africa will face significant increases in energy demand in the decades ahead, there is a clear need to consider how to meet the pressing energy needs of the continent.

In seeking to improve the supply of energy there are several major challenges. To date, energy policy in Africa has not succeeded in providing affordable, reliable and environmentally acceptable energy to all of the population. This failure has occurred despite the availability of significant indigenous hydrocarbon and renewable resources across the continent. In addition, many African countries lack the basic infrastructure that is essential for development. The financial and political situations are often difficult, thereby undermining the ability of local authorities to invest in the construction of the necessary infrastructure.

To date, private investment in Africa has concentrated on the exploitation of natural resources, notably hydrocarbons, for export to global markets rather than to satisfy local demand. The tendency of foreign investors to focus on exports of energy resources from producer countries arises from the fact that most local energy markets are not profitable. As a result, relatively little of the energy from producer nations is available to their neighbours.

Electricity offers one of the most effective means to promote sustainable development and to overcome the widespread energy poverty, especially in sub-Saharan Africa. Africa, however, is confronted with chronic underinvestment in the generation and transmission of electricity. The installed generation capacity in Africa is no more than that of Spain, and half of it is in South Africa. The continent lacks the capacity to satisfy current demand, let alone to meet future projections.

The challenge of extending electrification is not confined to the lack of investment, however. Africa also lacks the necessary legal and policy frameworks for developing the complex, cross-border networks needed to deliver electrical power to markets. Up to now, most energy projects have been implemented on the basis of bilateral agreements, normally involving a producer country and a multinational energy company or consumer country overseas. But bilateral arrangements do not provide the comprehensive framework necessary for stable regional cooperation on energy issues. There is therefore a clear role for international organisations, particularly those concerned with energy, to assist local authorities in rectifying the current situation.

2 The Role of International Energy Organisations

Various international organisations, particularly those concerned with development issues and finance, have an established presence in Africa. In recent years, these organisations have increasingly focused on the issue of energy, demonstrating their recognition of its importance for sustainable development. The World Bank, for example, is scaling up its energy investment programme in Africa, including increased support for trans-African or regional networks such as the West and Southern Africa Power Pools and interconnected national power grids. The African Development Bank is also making important contributions to improving the provision of energy in Africa.

There may be few international organisations dealing with energy, but they are decisive in setting the rules that are essential for an organised and civilised way of developing energy trade and infrastructure. While African countries have a significant involvement with a variety of international organisations, few of them are active members of the leading international energy institutions. In fact, the vast majority of African countries do not qualify for membership, even if they would wish to.

African countries are not members of the International Energy Agency, the prominent consumer organisation, which sets high barriers to membership. Africa's involvement in the Organisation of the Petroleum Exporting Countries (OPEC), the prominent supplier organisation, is limited to just four producer countries – Algeria, Angola, Libya and Nigeria.

Just two international energy organisations are open to African nations: the International Energy Forum (IEF) and the Energy Charter Treaty (ECT). The International Energy Forum is an important mechanism for dialogue that in recent years has begun to consider the provision of energy to the world's poor. The IEF is a meeting place where African countries can explain their situation and seek assistance. But because it can not enter into legally binding commitments, or offer immediate benefits, the IEF has significant limitations.

3 The Energy Charter Treaty

The Energy Charter Treaty (ECT) is of a different nature. Being the only multilateral treaty in the field of energy, it can be regarded by African countries as a tool that is particularly relevant for them. By signing and ratifying the Treaty, a country sends a powerful signal to the international community that it is willing to act in accordance with international rules that are already shared by more than 50 members, including the European Union and its member states, many Eurasian nations, as well as Japan, Turkey and others.

In terms of addressing the pressing need to tackle Africa's energy poverty, membership of the ECT offers a number of distinct advantages, notably with respect to investment promotion. Addressing Africa's chronic power shortages will require major investments in the refurbishment and expansion of power infrastructures. Much of the continent's currently installed electricity generation capacity is in need of refurbishment. Additional capacity will need to be built to meet suppressed demand, keep pace with projected economic growth and provide additional capacity to support the rollout of electrification.

One of the biggest challenges in overcoming energy poverty in Africa will thus be to mobilise the investment necessary to close the financing gap of more than US$30 billion per annum. Membership of the ECT gives a positive message to potential investors in energy projects who can be reassured by the security offered by the Treaty's provisions on investment protection.

Another key advantage of the ECT is the opportunity it provides to promote cooperative solutions appropriate to local and regional markets and demands. At present, 21 of the 48 sub-Saharan countries have an electricity generation capacity of less than 200 MW each. This is well below the minimum threshold necessary to achieve economies of scale. At the same time, the region has substantial energy potential – particularly in hydropower – but this is largely concentrated in a handful of countries that are far from centres of demand and lack the resources they need to expand their electricity generation capacity. In this situation, strengthened cooperation between African countries in the electricity sector would offer significant technical and commercial advantages.

The best way to expand generation capacity in Africa at the lowest unit cost is to develop large-scale generation projects to supply regional power pools through cross-border trade. The economic rationale for such projects is generally strong if the power they produce is intended to meet domestic needs in the countries where they are located, and foreign (export) needs in neighbouring countries. The establishment in Africa of large energy markets bound by shared rules would promote economies of scale in energy projects and the optimisation of transmission networks, and would offer increased opportunities to employ environmentally 'clean' sources – notably hydropower and wind. Together, these elements would lead to increased efficiency, reliability and common energy security.

As a good example of the development of such an approach, the Energy Charter Secretariat has recently supported the creation and operation of the Task Force on Regional Energy Cooperation. The task force brings together delegates from the countries of the region, representatives of international organisations and energy experts. The common aim of the group is to find effective and mutually acceptable ways to enhance cross-border trade in electricity and to create a regional electricity market.

The advantages of a regional approach to the supply of electricity, in particular, are clear in technical, commercial, but also in political terms. A regional approach would benefit from a common legal framework to guarantee investments and to ensure the development of the cross-border trade in electricity. Such rules are the basis of the Energy Charter and these can be readily adapted to the specific requirements of African countries.

3.1 The Energy Charter Process

Establishing practical mechanisms for cooperation in the energy sphere has other benefits. The creation of regular dialogue mechanisms, for example, can be an effective way to enhance transparency and to build trust and confidence among actors. On the basis of the ECT, member states have established the Energy Charter Process (ECP), which is designed to provide member states with opportunities to work cooperatively to advance the implementation of the Treaty, and to develop common responses to emerging issues.

The ECP could serve as a means to assist policy development and energy governance in developing countries, drawing on the experience and expertise of member states and of the Secretariat. The Charter supports the analysis of trade, transit issues, transit tariffs and access conditions. It also deals with issues related to trade distortions and trade facilitation, and encourages interactions with the private sector through a dedicated forum.

The Energy Charter Process, while remaining true to the Treaty and its principles, is a dynamic and evolving community. Membership of the ECT allows states to participate in the further development of the Treaty and its associated institutions to meet new challenges. The ECP is a very effective way to join a broad community, to explore specific and common problems and to learn and share solutions. In today's globalised world, it is an advantage not to be isolated. The Treaty is open to all states that are ready to share its basic principles, the first and perhaps most important for Africa being sovereignty over energy resources.

4 Conclusion

One way to ensure that globalisation works to the advantage of all peoples is to engage the main actors in the energy sector through international energy organisations. Africa cannot escape globalisation, certainly not in the field of energy. International experience has highlighted the key role that multilateral cooperation between states can play in promoting mutually beneficial energy policies. Unfortunately, most African countries are not members of the few international organisations that specialise in energy. Given this, African governments should be encouraged to adopt a policy of more active participation in institutions such as the ECT, which would offer them numerous advantages.

The integration of Africa's energy markets by enhancing the environment for investments, together with steps to ensure the reliable and secure transit of energy products, would do much to improve the provision of energy. The advantages of a regional approach to the supply of electricity, in particular, are clear – in technical, commercial, but also in political terms. A regional approach would benefit from a common legal framework to guarantee investments, and would ensure the development of cross-border trade in electricity. Such rules are the basis of the Energy Charter and can be readily adapted to the specific requirements of African countries.

Nevertheless, African countries cannot assume that the development of energy trade and infrastructure can be secured simply by becoming a member of an international energy organisation. The actions of specialised financial institutions and others are also key to success. The World Bank and the African Development Bank, NGOs that offer bilateral assistance and donor states all have vital roles to play. Global, regional and local partnerships will be vital, as will be building upon the initiatives already under way in this area.

The G8 energy ministers' meeting in Rome in 2009 recognised the importance of broad international cooperation to tackle Africa's energy poverty. At that meeting, the ministers invited the Energy Charter Treaty, among other organisations, to engage African and international institutions in the development of a strategy to combat the continent's energy problems. Such cooperation with other key players should ensure that this initiative makes a substantial contribution to reducing and eventually ending energy poverty in Africa.

Resolving Mismatches in Energy Decision Making

Richard R. Cirillo

Abstract This chapter discusses the process by which decisions related to the energy sector are made. The complexity of the energy system makes energy decisions difficult. The system involves an interdependent set of supply technologies such as coal, oil, natural gas, nuclear, electricity and renewables, and demand sectors, including the residential, commercial, industrial, agriculture and transportation sectors, each with their own special energy requirements. The energy system is dynamic and changes quickly. Energy decision makers need to look back and learn from past experiences, including the outcomes of previous energy policies and projects. They must look ahead and anticipate what might happen in the future. They also need to look sideways and consider both supply-side and demand-side components, non-energy infrastructures such as transportation, water supply and communication, and the environment.

1 Introduction

This chapter discusses the process by which decisions related to the energy sector are made. Energy is critical for economic development, alleviating poverty, promoting public health and maintaining a clean environment. All countries are working to make good choices about their future supplies of energy with considerations of cost, security, reliability, environmental protection, climate change and technology developments.

R.R. Cirillo (✉)
Decision and Information Sciences Division, Argonne National Laboratory,
Argonne, IL 60439, USA
e-mail: cirillor@anl.gov

F.L. Toth (ed.), *Energy for Development: Resources, Technologies, Environment*,
Environment & Policy 54, DOI 10.1007/978-94-007-4162-1_6,

First, Sect. 2 describes the challenges involved in energy decision making. Section 3 provides a general framework for decision making, and Sect. 4 uses four examples of energy decision making in the United States to illustrate how the choices were made and what their implications were. Finally, Sect. 5 presents a summary of how energy decision making should be conducted.

2 Challenges in Energy Decisions

Energy decision makers include government policy makers and regulators, energy company officers from both the public and private sectors, and energy users. They are regularly confronted with complex and difficult choices. These can involve energy policies, energy supply projects, end-use technologies, energy efficiency, environmental protection, energy security and international agreements, and an entire range of other complicated issues.

It is difficult to make energy decisions because the energy system is complex. It is an interdependent set of various supply technologies – coal, oil, natural gas, nuclear, electricity and renewables. It is equally complex on the demand side, including the residential, commercial, industrial, agriculture and transportation sectors, each with their own special energy requirements. The energy system is dynamic and changes quickly. A good example of these characteristics is the fluctuation in the price of oil on the international markets from day to day. The energy system also involves many uncertainties. Questions abound as to the future demand for energy, the availability of energy resources, the price of energy and how new technologies will perform.

Energy decisions have significant implications, as seen in other chapters (see Difiglio 2012; Gibbons 2012; Gupta 2012). They are a critical component of development and will have a significant impact on the development process of all countries. Often, it takes a long time to make energy decisions, and these decisions have long-lasting effects. Many components of the energy system will be in place for 30, 40, 50 or more years. Given these considerations, it is important to understand how decisions are made in the energy system and how they can be made more effectively.

Energy decisions rely heavily on information provided by energy analysts. These analysts conduct trend analyses by looking at historical information. They conduct optimisation analyses to attempt to maximise or minimise some objective function. They perform equilibrium analyses to find how the variations in the energy system may gravitate to some equilibrium state. They conduct sensitivity analyses and uncertainty studies to better understand the potential range of possible future outcomes.

Good analysis is always important in making good decisions. However, analysts and decision makers need to recognise the concept of ‘bounded rationality’ (Simon 1991a). This concept is based on the recognition that individuals make decisions and choices that are constrained by the amount and quality of information they have available, by their own mental ability to receive and process that information, and by the time constraints imposed on coming to a decision. Energy decision makers, no matter how smart and how well intentioned they are, are often forced to make choices under constrained circumstances. Often, they have limited, incomplete, and sometimes unreliable information upon which they have to base a decision. Sometimes they face the opposite situation – they have too much information and a limited ability to absorb and process it all. Further, it is not uncommon for there to be a limited period of time in which energy decisions must be made, and it is not possible to wait for a detailed analysis to be completed.

Decision makers in general and energy decision makers in particular are not optimisers that always select the best possible path. They are referred to as 'satisficers', a term coined by Herbert Simon (1991b), who was awarded the 1978 Nobel Prize in economics. A satisficer makes decisions that are often suboptimal. They are not the absolute best decisions, but hopefully they are satisfactory or good enough for the time and conditions at hand. Given the fact that energy decision makers must be satisficers and not optimisers, the ability to adapt and adjust is an important, if not critical, part of good energy decision making.

3 A General Framework for Decision Making

A general description of the energy decision-making process is presented in Fig. 1. First, energy decision makers need to look back – they need to learn from past experiences, not only in their own country but also in other countries. This look back involves reviewing historical trends in energy supply and demand. In addition to a country's own data, several international organisations provide compilations of

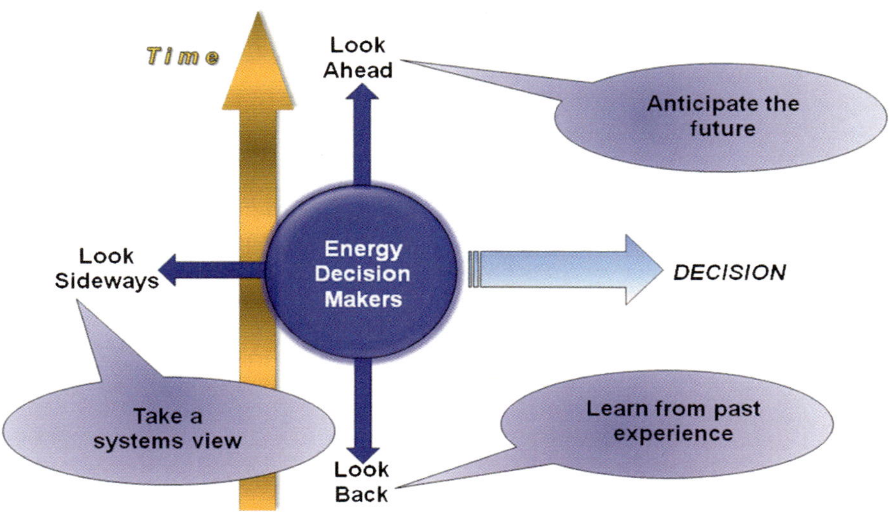

Fig. 1 Generalised energy decision-making process

data from many countries. The International Energy Agency (IEA 2010a) and the International Atomic Energy Agency (IAEA 2010) are examples of sources of this information.

In addition to reviewing statistical data, the look back involves a review of previous energy policies and projects, and the factors that contributed to their success or failure. While there is no single source for this type of information, many technical and popular publications have documented experiences in energy decision making (Rivera and Vargas 2007; Yue and Yang 2007; MacGill 2010; Wang et al. 2010).

Second, energy decision makers must look ahead and anticipate what might happen in the future. They must do this with awareness of the wide range of uncertainty in future developments. Several international organisations and research institutes publish projections of future energy supply and demand based on simulation models and other analyses (WEC 2007; EIA 2010a; IEA 2010b). In addition, a multitude of published material is available on the status of emerging energy technologies, projections of energy prices and other issues that could affect future energy development.

Third, energy decision makers need to look sideways. When looking at energy, decision makers must view it as part of a large, complex and interdependent system. The energy system includes both supply-side and demand-side components. Decisions made in one area will affect many other areas. Further, the energy system does not operate independently. It depends on other infrastructures, such as transportation, water supply and communications. It is part of the larger national economy, part of a larger global economy and part of the biological ecosystem that sustains life. It is vitally important that energy decisions include recognition of these interdependencies.

Once an energy decision maker has looked back, looked ahead and looked sideways, an informed decision can be made.

4 Examples of Energy Decision Making in the United States

The following subsections present examples of energy decision making in the United States. These descriptions are not intended to criticise the decisions or to pass judgment on the decision makers. Rather, the intent is to focus on what factors went into each decision and on what information the decision makers based their choices.

4.1 Example 1: Natural Gas Power Plant Construction

Natural-gas-fired power plant construction in the United States is shown in Fig. 2. A sizeable peak in construction of these plants occurred in the period 2000–2004, and then construction levelled off. Several factors contributed to the decisions to

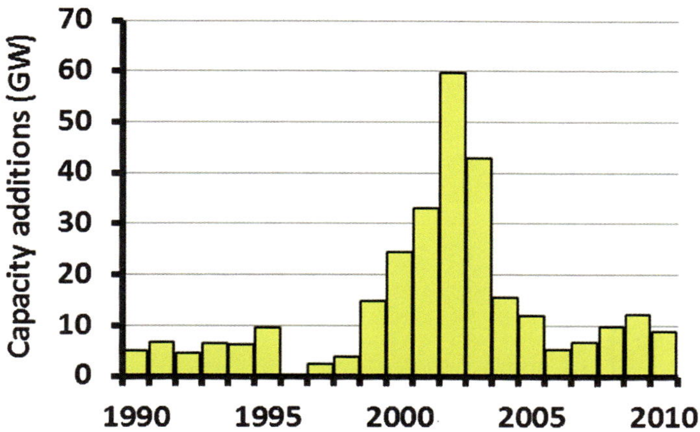

Fig. 2 US natural gas generation capacity additions, 1990–2010 (Source of data: EIA 2010b)

carry out this unusual rate of construction. Figure 3 illustrates some of the information that was available to decision makers.

First, when looking back, decision makers saw a significant increase in the electrical load in the country, indicated by the blue line in Fig. 3a. They also saw limited capacity additions resulting in reserve margins dropping from about 22% to less than 15% in the 8-year period 1990–1998, shown by the red line in Fig. 3a. Further, they saw that natural gas prices had been stable and relatively low for the better part of a decade, as shown in Fig. 3b.

In looking ahead, energy decision makers knew that the short construction time for natural gas units, typically 2–3 years, would result in a rapid payback on investment. In addition, electricity deregulation was coming into play in the United States. This change meant that companies would have ample opportunity to build profitable businesses around electricity generation. These factors led to decisions to build many natural gas generation units quickly.

Unfortunately, the sideways look did not receive enough attention. The increased demand for natural gas for electricity generation caused, among other things, natural gas prices to rise considerably – by a factor of approximately 3 over this period, as shown in Fig. 4. Further, the State of California had an adverse experience with an electricity deregulation programme that did not work well. As a result, the construction boom resulted in significant generation overcapacity. Many units were idled for years and lost money for their investors over this period.

4.2 *Example 2: Transportation Sector*

Transportation energy demand is an issue that has received considerable attention in the United States. There is a need to reduce the transportation sector's oil demand and the carbon dioxide (CO_2) emissions from vehicles. Figure 5 shows the growth

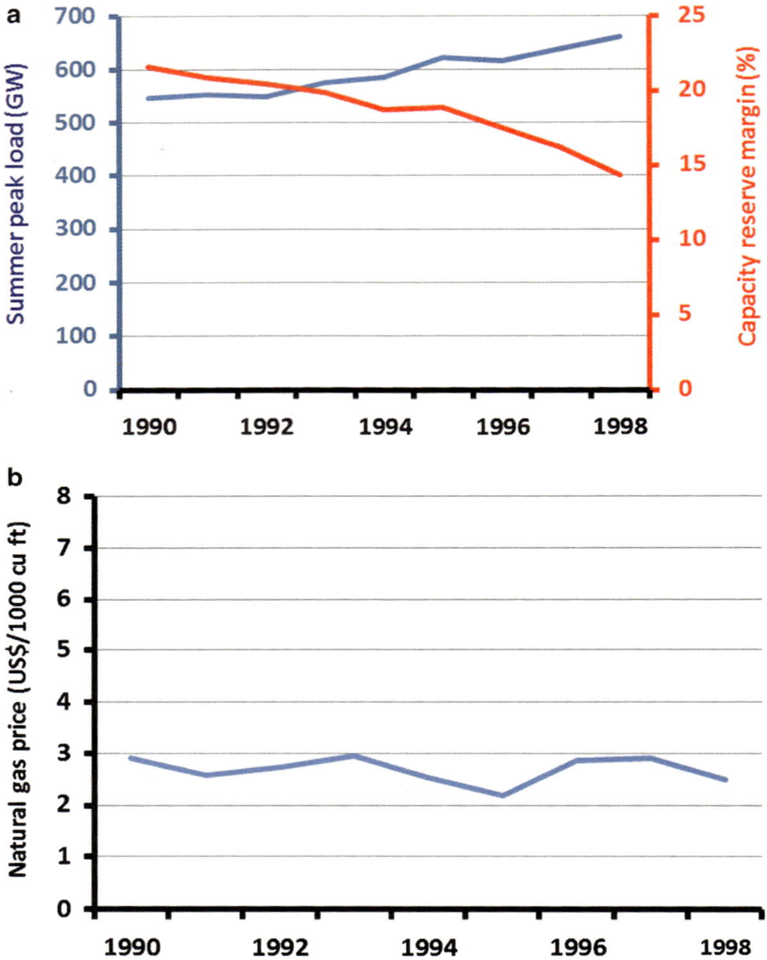

Fig. 3 US peak load and natural gas prices, 1990–1998 (Source of data: EIA 2010b). (**a**) US Peak electricity load and reserve margin, 1990–1998 (Source of data: EIA 2010b). (**b**) US natural gas prices 1990–1998 (Source of data: EIA 2010b). Note: *cu ft* cubic feet (1 cu ft = 0.0283 m^3)

in gasoline consumption in the United States dating back to 1980. The small dip at the end is a result of the current global recession. There is concern that when the recession is over, the same growth trend will continue.

Identifying the factors that contributed to this situation starts with a look back. Gasoline prices in the United States have historically been very low, as shown in Fig. 6a (note that prices are expressed in inflation-adjusted constant US dollars). After the oil price shocks of the 1970s, there was a steep decrease in gasoline prices, and they remained flat for nearly two decades. Even the price increases over the last

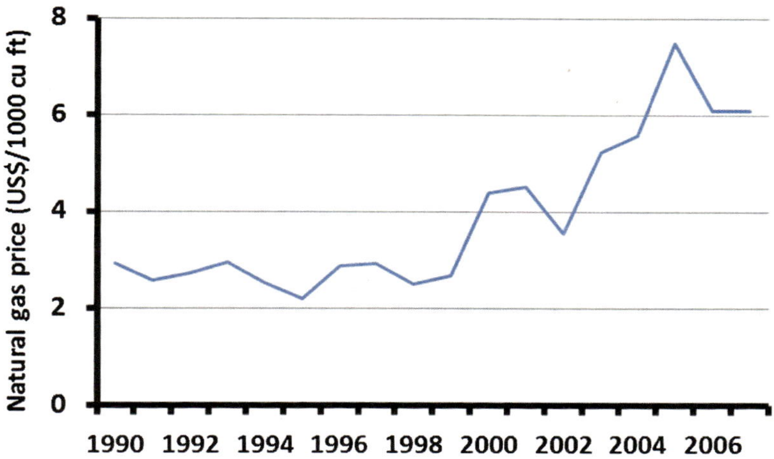

Fig. 4 US natural gas prices, 1990–2007 (Source of data: EIA 2010b; Note: *cu ft* cubic feet (1 cu ft = 0.0283 m^3))

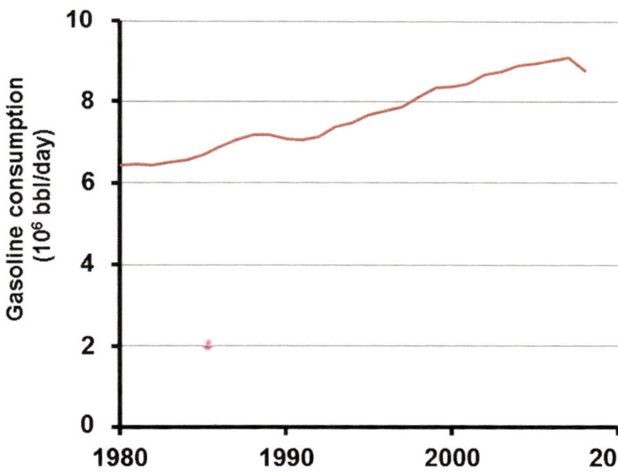

Fig. 5 US gasoline consumption, 1990–2010 (Source of data: EIA 2010b; Note: *bbl* barrel (1 bbl = 0.159 m^3))

5 years have been relatively modest when expressed in constant currency. During the same period, there has been a significant increase in the use of vehicles. The blue line in Fig. 6b shows that the number of registered vehicles in the United States has grown significantly over this period. The red line shows that travel, expressed in vehicle-miles travelled, has almost doubled in this same period. In addition, the types of vehicle being used have changed, as shown in Fig. 6c. In 1980, about 80% of the new vehicles purchased were sedans, which are the commonly known

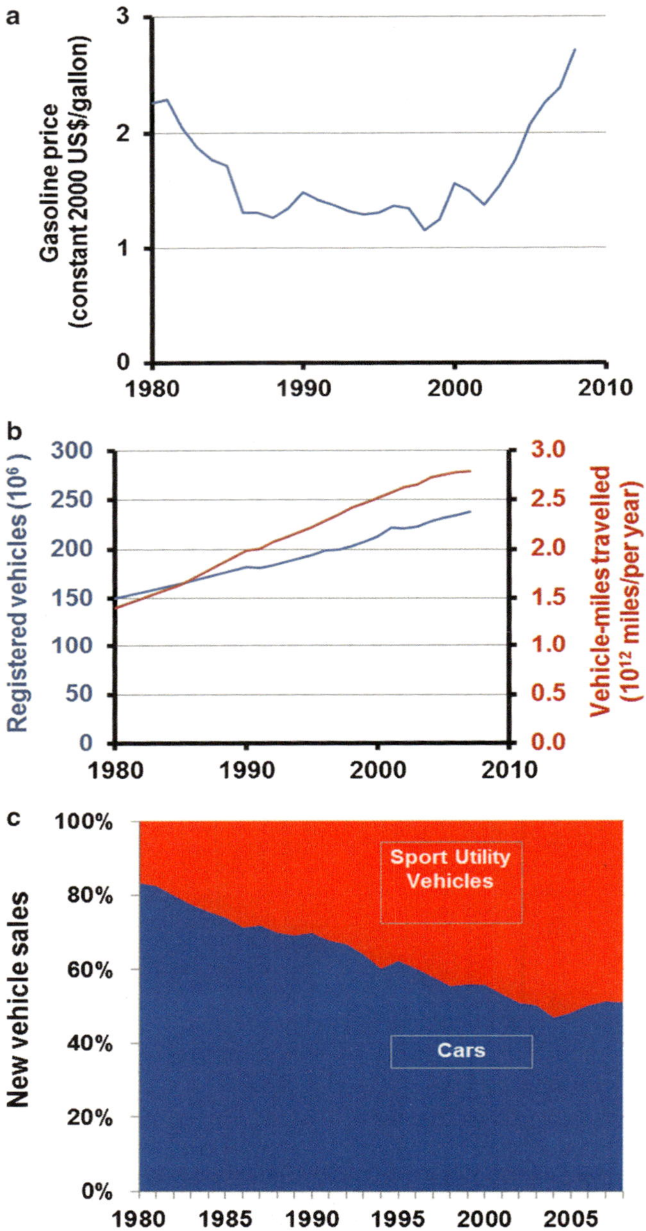

Fig. 6 Trends in the US transportation sector, 1980–2010. (**a**) US Gasoline prices (Source of data: EIA 2010b). (**b**) US Vehicles registered and vehicle-miles travelled (Source: RITA 2009). (**c**) US New vehicle sales by type (Source: RITA 2009). Note: 1 mile = 1.6093 km

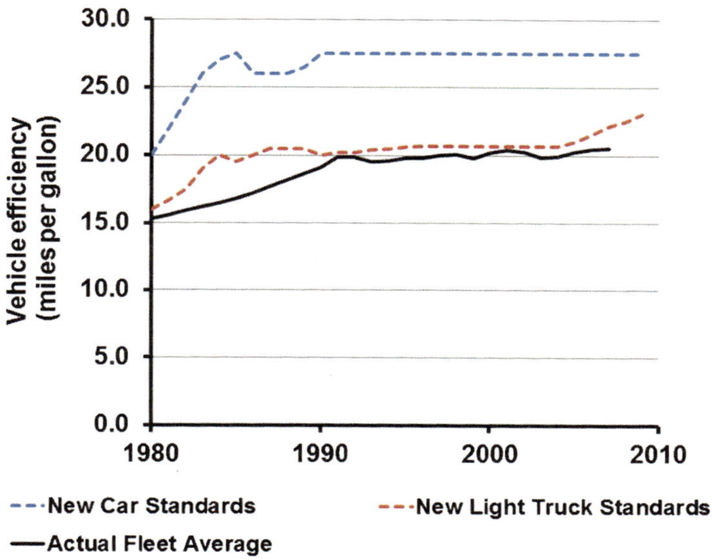

Fig. 7 US vehicle efficiency standards and actual fleet efficiency, 1980–2010 (Sources of data: EIA 2010b; NHTSA 2010; Note: 1 mile/gal = 0.425178 km/l; 15 miles/gal corresponds to about 15.68 l/100 km; 27 miles/gal corresponds to about 8.71 l/100 km)

automobiles or cars. By the mid-2000s, the commonly used vehicle had shifted – about half the new vehicles purchased were sport utility vehicles, which are built on the same frame as a small truck. Typically, these vehicles are significantly heavier and have much poorer fuel efficiency.

Decision makers are currently confronted with several options to deal with this situation. The first is to increase vehicle efficiency standards. In the past, this option has been successful but has been affected by the vehicle mix and the turnover rates, as shown in Fig. 7. The top line shows the vehicle efficiency standards for new automobiles. The middle line shows the efficiency standards for new light-duty trucks or sport utility vehicles. The bottom line shows the average vehicle efficiency of the whole fleet, including both new and existing vehicles. It is evident that the standards have resulted in efficiency improvements since 1980. However, the 30% improvement in efficiency has not been enough to compensate for the doubling of vehicle-miles travelled. It should be noted that the efficiency standards for both types of vehicle are scheduled to increase from the current standard of about 27 miles per gallon (about 8.71 l/100 km) to about 35 miles per gallon (about 6.72 l/100 km) by 2020.

Another option is to develop alternative fuels to reduce the dependency on oil. In the United States, the primary alternative fuel currently being utilised is ethanol. As a result of this emphasis, the capacity to produce ethanol has grown rapidly, as shown in Fig. 8, experiencing more than a sixfold increase in a 10-year period. In the United States, the primary source of ethanol is corn (maize), a widely produced crop.

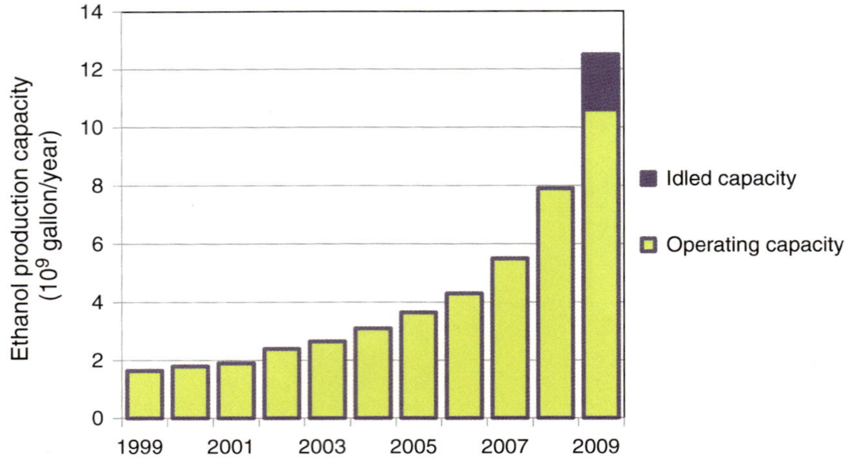

Fig. 8 US ethanol production capacity, 1999–2009 (Source: RFA 2011; Note: 1 gal = 3.785 l)

To understand the impact of increased ethanol production on the demand for and price of corn, it is important to look sideways at the larger system that includes corn. In addition to being used for ethanol production, corn is widely used as a food source and for animal feed; it is also exported. The increase in demand for corn to be used in ethanol production resulted in significantly higher corn prices. The higher price of corn, combined with the drop in crude oil prices from the peak of US$140 per barrel down to their current levels, resulted in ethanol losing its competitive advantage against gasoline. The result has been idled ethanol production capacity and a large number of company bankruptcies. The purple bar in Fig. 8 shows that about 15% of the ethanol production capacity has been idled because of the drop in demand.

Research is underway in the United States on cellulosic ethanol production, which uses materials other than corn as feedstocks. Crops such as switchgrass, which can be grown on otherwise non-productive agricultural land, are being tested. These alternatives, however, are still in the research stage and are not economically competitive at this time.

4.3 Example 3: Advanced Vehicle Technology

Another option for consideration by decision makers is the development of advanced vehicle technology. The technology that is attracting growing interest in the United States is called the 'plug-in hybrid.' This is a vehicle that has a large battery pack along with a gasoline engine. The range of the vehicle on batteries alone is about 40 miles (64.4 km) on a single battery charge. When using both the batteries and the gasoline engine, the plug-in hybrid can travel about 300 miles (483 km), which is

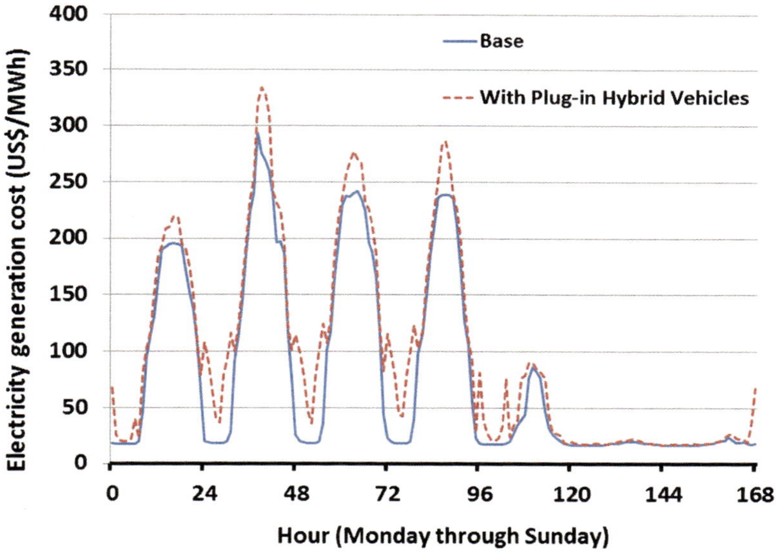

Fig. 9 Effect of plug-in hybrid vehicles on electricity generation costs in a region of the United States (Source: ANL 2009)

about the same as a conventional automobile. The vehicle is expected to become available in 2011, but its cost has not yet been announced. The announced efficiency (Chevrolet 2011) of the plug-in hybrid is the equivalent of 230 miles (370 km) per gallon (1.02 l/100 km), which is about ten times better than any vehicle currently available. It should be noted, however, that this estimate is based on a different type of driving cycle. Nevertheless, it is a significant improvement in fuel efficiency over current automobiles.

Again, it is important to look sideways at the system issues. Figure 9 shows the results of a study performed at Argonne National Laboratory on the impact of plug-in hybrid vehicles on the electric power grid. The graph, for a particular region in the United States, shows the hourly costs of electricity generation over 1 week. These costs consider the effect of transmission grid congestion on the selection of which generating units to run. The solid line shows the baseline case without the plug-in hybrids. The dashed line shows the effect of a modest penetration of plug-in hybrids. As many people plug-in at night to charge vehicle batteries, there is an increase in the load and also in electricity generation costs. There is a beneficial night-time load-levelling effect, but there is also the potential for a significant increase in the daytime peak load, as some individuals will attempt to charge batteries during the day. This increase could severely stress the power grid when it may already be at or near maximum load conditions. It would also result in an increase in daytime electricity generation costs and potentially consumer prices.

It has been suggested that the use of plug-in hybrid vehicles will lower CO_2 emissions by displacing some of the need to burn gasoline because a portion of the travel

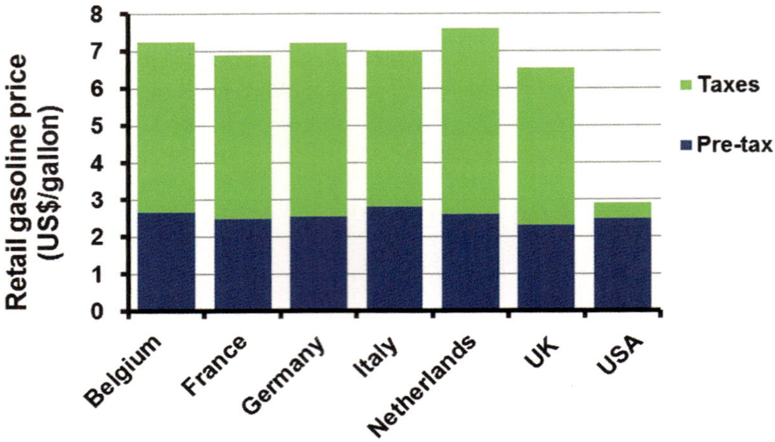

Fig. 10 Comparison of gasoline prices in various countries, 2011 (Source: IEA 2011; Note: 1 gal = 3.785 l)

will be using electric batteries. However, CO_2 emissions may increase or decrease depending on the mix of fuels used for power generation. At this stage in the development of this advanced technology, the results are uncertain.

Yet another option is to impose higher taxes on gasoline to encourage the use of more efficient vehicles. Figure 10 shows a comparison of the cost of gasoline in the United States and in six European countries – Belgium, France, Germany, Italy, the Netherlands and United Kingdom. The pre-tax price is more or less the same across Europe and the United States; however, the tax levels are very different. European countries have levied a much heavier tax load on gasoline than the United States. The tax in Europe is more than the price of the gasoline itself. A higher tax on gasoline might promote more efficient vehicles, but a look sideways reveals a more complex situation. The US economy is much more heavily dependent on motor vehicles, both cars and trucks, for transportation than are the European economies. A significantly higher gasoline tax would have substantial economic impacts. In addition, taxes are not a popular mechanism to achieve objectives.

4.4 Example 4: Nuclear Power

Figure 11 shows nuclear power capacity additions in the United States from 1960 to the present. Two significant peaks occurred, one in the mid-1970s and one in the mid-1980s. The question arises as to what was in decision-makers' minds that resulted in this unusual pattern. First, it is important to note that in the United States virtually the entire electric power system is privately owned; capacity expansion decisions are made by private sector companies, not the government. Second, the sharp drops from the peak capacity addition levels coincided with the accidents at

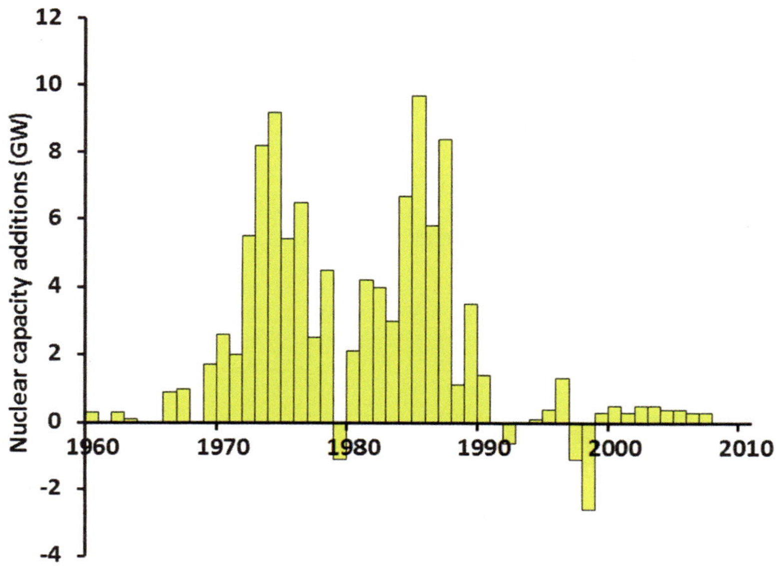

Fig. 11 US nuclear capacity additions, 1960–2010 (Source of data: EIA 2010b)

the Three Mile Island nuclear plant in the United States in 1979 and the Chernobyl nuclear plant in what was then the Ukrainian Soviet Socialist Republic in 1986. Those accidents created a lot of public opposition to nuclear power. Hirschberg (2012) discusses the effects on the externality costs of nuclear power projects. Because nuclear units have a high up-front capital cost, financing organisations became wary of providing substantial loans to projects that might face serious public opposition. Further complicating the decisions on proceeding with nuclear power development were issues of nuclear waste management and weapons proliferation. As a result of this looking back, decisions were made by the private utilities not to pursue nuclear power in the United States.

A look ahead is needed to see what the future might hold. In the short term, the issues that led to the halt in nuclear plant construction – financing, waste management and weapons proliferation – have not gone away. In the longer term, new issues have arisen that may influence decisions: profitability and climate change. A news report in April 2009 stated, 'Exelon … said that its first quarter earnings rose 23%, driven by increased output at its nuclear operations and higher rates' (Williams 2009). Exelon is the largest nuclear utility in the United States. Its nuclear power plant operations produced a significant growth in earnings in the middle of a US recession.

In another news article in May 2008, the *Wall Street Journal* reported, 'As [the US] Congress debates whether to limit carbon-dioxide emissions … the nuclear power industry is poised to reap a multibillion-dollar windfall if restrictions take effect' (Smith 2008). The combination of profitability arising from the normally low operating costs of nuclear plants and the potential for significant increases in the operating costs of fossil fuel plants due to CO_2 emission restrictions may make nuclear power much more attractive to decision makers. At this time, decisions on the future of nuclear power in the United States have not yet been made. There continues to be significant uncertainty and much analysis is still being conducted.

5 Summary and Conclusions

The following points summarise some basic principles for energy decision making:

- *Recognise that the energy system has no simple solutions, only choices.*
  Henry Mencken, an American essayist, observed that 'For every complex problem there is a solution that is simple, neat – and wrong'. Energy decisions are inherently complex. Simple solutions may sound attractive but can lead to very poor results. In general, choices have to be made between options that have both positive and negative aspects.
- *Look back and learn from past experience.*
  It is important to learn from past experiences both in one's own country and in other countries. This look back helps to avoid repeating mistakes that others have made.
- *Look ahead to evaluate the future and don't be near-sighted.*
  Predicting the energy future is always difficult because of the complex uncertainties involved. Nevertheless, it is necessary to try to anticipate future developments. In doing so, it is important not to be near-sighted and focus only on short-term problems. Long-term planning can help keep tomorrow's problems from becoming today's problems.
- *Look sideways and treat energy decisions as systems decisions.*
  Whatever decision is made in any part of the energy system will undoubtedly affect some other part of the system; some part of the national, regional or global economy; some part of an ecological system; and/or some social system. These effects are often difficult to understand and predict. Nevertheless, they need to be part of the decision-making process.
- *Make energy decisions that are adaptable and adjustable.*
  Albert Einstein noted that 'Insanity is doing the same thing over and over again and expecting different results'. Energy decisions need to be adaptable and adjustable. As conditions change, it is important to learn from mistakes and make better decisions. Adaptation and adjustment have been the secret of survival in the biological world; it is the secret of survival in the energy world as well.
- *Don't let the 'optimum' get in the way of the 'good enough'.*
  Often in energy decision making, there is a struggle to find the absolute best energy situation. Decisions are delayed and deferred while waiting for a thorough analysis. In many cases, avoiding a crisis requires making a decision that may not be perfect but is good enough. Improvements can be made later.

- *Make decisions using good data and good analysis.*
  Will Rogers, an American humourist said, 'It isn't what we don't know that gives us trouble, it's what we know that isn't so'. It is important to make good energy decisions using the best available data and the best possible analysis to improve what we do know and to enhance what we understand.

If all or some of these principles are followed, the energy systems in all countries will be better off.

References

ANL (Argonne National Laboratory). (2009, November 6). *Impact of plug-in hybrid electric vehicles on the electricity market in Illinois.* News. www.dis.anl.gov/news/Illinois_PluginHybrids.html. Accessed 13 May 2011.

Chevrolet. (2011). *Chevrolet information*. www.chevrolet.com. Accessed 11 May 2011.

Difiglio, C. (2012). Financing power sector investments. In F. L. Toth (Ed.), *Energy for development: Resources, technologies, environment* (pp. 161–170). Dordrecht: Springer.

EIA (US Energy Information Administration). (2010a). *International energy outlook 2010.* Washington, DC: US Department of Energy.

EIA (US Energy Information Administration). (2010b). *Annual energy review, DOE/EIA-0384.* Washington, DC: EIA.

Gibbons, J. H. (2012). Technology and innovation. In F. L. Toth (Ed.), *Energy for development: Resources, technologies, environment* (pp. 141–148). Dordrecht: Springer.

Gupta, S. (2012). Financing renewable energy. In F. L. Toth (Ed.), *Energy for development: Resources, technologies, environment* (pp. 171–186). Dordrecht: Springer.

Hirschberg, S. (2012). Externalities in the global energy system. In F. L. Toth (Ed.), *Energy for development: Resources, technologies, environment* (pp. 121–138). Dordrecht: Springer.

IAEA (International Atomic Energy Agency). (2010). *Country nuclear power profile, 2010 edition.* Vienna: IAEA.

IEA (International Energy Agency). (2010a). *Energy statistics manual.* Paris: OECD/IEA.

IEA (International Energy Agency). (2010b). *World energy outlook 2010.* Paris: OECD/IEA.

IEA (International Energy Agency). (2011). *End-use petroleum product prices and average crude oil import costs.* Paris: OECD/IEA.

MacGill, I. (2010). Electricity market design for facilitating the integration of wind energy: Experience and prospects with the Australian national electricity market. *Energy Policy, 38*(7), 3180–3191.

NHTSA (National Highway Transportation Safety Administration). (2010). *Summary of fuel economy performance.* Washington, DC: US Department of Transportation/NHTSA.

RFA (Renewable Fuels Association). (2011). *Ethanol industry outlook 2011: Building bridges to a more sustainable future.* Washington, DC: RFA.

RITA (Research and Innovative Technology Administration). (2009). *National transportation statistics 2009.* Washington, DC: US Department of Transportation/RITA/Bureau of Transportation Statistics.

Rivera, S. R., & Vargas, A. (2007). Methodological basis to decide short-term investments in distribution systems under uncertainty in performance-based regulatory frameworks. *International Journal of Global Energy Issues, 28*(1), 88–110.

Simon, H. A. (1991a). Bounded rationality and organizational learning. *Organization Science, 2*(1), 125–134.

Simon, H. A. (1991b). *Models of my life.* New York: Basic Books.

Smith, R. (2008, May 19). Carbon caps may give nuclear power a lift. *Wall Street Journal*. http://online.wsj.com/article/SB121115810973702419.html?mod=hpp_us_whats_news#articleTabs%3Darticle. Accessed 17 May 2011.

Wang, B., Kocaoglu, D. F., Daim, T. U., & Yang, J. (2010). A decision model for energy resource selection in China. *Energy Policy, 38*(11), 7130–7141.

WEC (World Energy Council). (2007). *Energy policy scenarios to 2050*. London: WEC.

Williams, M. (2009, April 23). Exelon 1Q earnings up 23 percent on nuclear output. *Seattle Times*. http://seattletimes.nwsource.com/html/businesstechnology/2009108643_apusearnsexelon.html. Accessed 13 May 2011.

Yue, C.-D., & Yang, G. G.-L. (2007). Decision support system for exploiting local renewable energy sources: A case study of the Chigu area of southwestern Taiwan. *Energy Policy, 35*(1), 383–394.

Part II
Energy Demand

Demography, Urbanisation and Energy Demand

Shonali Pachauri

Abstract Urbanisation is a major demographic driver of energy demand. It is intertwined with many aspects of human behaviour, including living and working arrangements and the organisation of economic activity. This chapter explores the differences in energy access and use between rural and urban households in developing countries, particularly India and China. The analysis shows the importance of future patterns of migration, urbanisation and changes in urban density for both the household energy mix and the evolution of energy demand in the coming decades. The amounts and types of energy used by rural and urban households will continue to differ significantly. Concerted efforts and investments will be needed to improve the living conditions of low-income urban households and their access to more efficient and safer forms of energy.

1 Key Demographic Drivers of Energy Use

This chapter discusses the impacts of urbanisation, a major demographic driver, on energy demand. Urbanisation is intertwined with many aspects of human behaviour, including living and working arrangements and the organisation of economic activity, and so can affect energy demand in several ways (see Gruebler et al. forthcoming). This chapter focuses on the differences in energy access and use between rural and urban households in developing countries.

First, this section discusses some of the key demographic drivers of energy use, and the treatment of demographic changes in energy demand models. Section 2

S. Pachauri (✉)
Energy Program, International Institute for Applied Systems Analysis (IIASA), Schlossplatz 1, A-2361 Laxenburg, Austria
e-mail: pachauri@iiasa.ac.at

F.L. Toth (ed.), *Energy for Development: Resources, Technologies, Environment*, Environment & Policy 54, DOI 10.1007/978-94-007-4162-1_7,

presents recent trends in urbanisation and future projections for different world regions. Section 3 discusses urban households and their access to modern sources of energy. Section 4 describes the differences in energy access and use between rural and urban households in China and India. Section 5 presents scenarios of household energy use for regions where the lack of access is most acute. Finally, Sect. 6 presents some policy implications and suggests areas for further research.

Demographic changes, economic growth, technological shifts, policies and lifestyles are among the most commonly agreed drivers of energy use and greenhouse gas emissions (OECD 2001, 2007; WEC 2003; Winistorfer et al. 2005; Price et al. 2006; IEA 2009). Other chapters in this volume have discussed alternative schools of thought on how these factors influence energy use (Luna and Gomelsky 2012; Brew-Hammond 2012). But there has been no debate on the importance of demographic changes, economic growth and technology as drivers of energy demand (Reilly et al. 1987, 1999; Martin and Richels 1979; Prinn et al. 1999; Nakicenovic and Riahi 2001, 2003; Nakicenovic et al. 2006; Riahi et al. 2007).

Demographic factors play a key role in determining energy use (O'Neill et al. 2010). Population size matters, but so also do other variables such as household size and structure, and the age composition of the population (Birdsall 1992; O'Neill and Chen 2002; O'Neill 2003, 2005; Dalton et al. 2008). For instance, studies suggest that household size has a significant effect on energy consumption, as larger families can benefit from economies of scale (Pachauri 2004; Lenzen et al. 2004, 2006). Urban households are often significantly smaller than rural households because there are fewer children or extended family members living together under the same roof. Thus, with increasing urbanisation, one can expect a larger number of small households. Ageing also impacts energy use, reflecting changes in the ways people use their time. As people age and retire, they spend more time at home and so need to heat and light their homes for longer periods. At the same time, however, they do not need to commute to work, but might take more leisure trips, which will have an impact on their transport energy needs (Allen and Edmonds 1981; Ha et al. 1993; O'Neill and Chen 2002; Prskawetz et al. 2004; O'Neill 2005; Tonn and Eisenberg 2007).

As cities have grown, there has been an urbanisation of poverty and a growth of populations living in informal or slum settlements (Ravallion et al. 2007; Baker 2008; Kitio 2007). Slum growth represented 38% of overall urban growth between 2000 and 2010 (World Bank 2010). Despite urban populations, on average, having better access to energy infrastructures than rural populations, 15% of the world's urban population still lack access to electricity, and about 70% in developing countries use liquid fuels such as kerosene or bottled gas for cooking and heating (IEA 2010; UNDP/WHO 2009). In sub-Saharan Africa the situation is particularly alarming, with only 46% of the urban population using electricity and 42% using modern fuels. The concern is that continuing urbanisation might be accompanied by growing numbers of urban poor without access to modern forms of energy, and a deterioration in the quality and reliability of energy services available to them. Uncontrolled urbanisation, as is evident in many developing countries, exerts considerable pressure on land, housing, and social and physical infrastructure, leading to the spread of slums (Dhingra et al. 2008). Unlike housing, water, sanitation and roads, energy is not recognised as a basic service and is therefore not on the radar of the urban development authorities as an issue that demands attention. Without concerted action to improve access to modern forms of energy for the urban poor, slums may emerge as a serious concern in the coming decades.

In addition to the general trend of increasing urbanisation, the nature of urbanisation – whether in small towns or large metropolitan areas – will also have impacts for future energy access and use. Recent evidence indicates that poverty is both more widespread and deeper in very small and small towns than in large or very large cities (Ferré et al. 2010), and is generally compounded by greater deprivation in terms of access to basic infrastructure services, including energy. Thus, rapidly growing small towns and cities will need to pay particular attention to ensure the provision of basic services, including electricity and modern fuels, to their populations.

2 Urbanisation Trends and Projections

In the 1970s, about 36% of the world's population lived in urban areas. In 2008, for the first time, this number rose to about 50%, and by 2030 it is projected to increase to 60% or more (Fig. 1, left panel) (UN DESA 2007). A brief examination of urbanisation trends by region (Fig. 2) reveals a marked heterogeneity in the pattern of urbanisation between regions and variations in the level and speed of urbanisation. In most OECD countries and Latin America, urbanisation levels are already high, about 70% or more, so that future growth in urbanisation will be limited. In the coming decades, most urban growth will occur in areas where levels of urbanisation are low but rising fast, mainly in Asia and parts of Africa. By 2030, more than half of the world's urban population will reside in Asia, while Europe's share will fall from 38% in 1950 to just 11% in 2030.

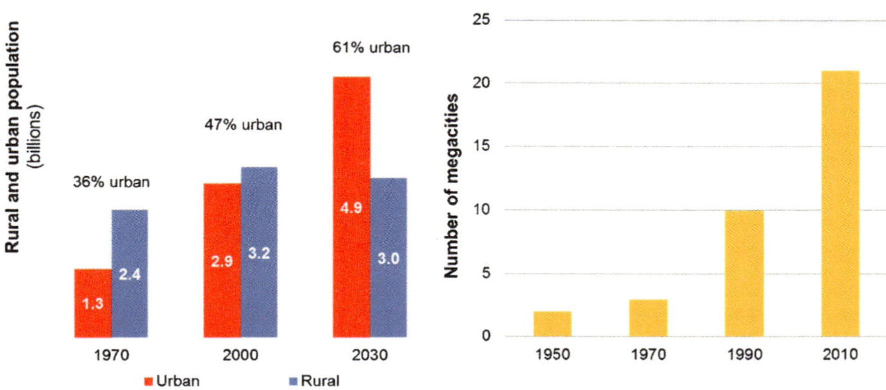

Fig. 1 World urbanisation trends (Source: UN DESA 2007)

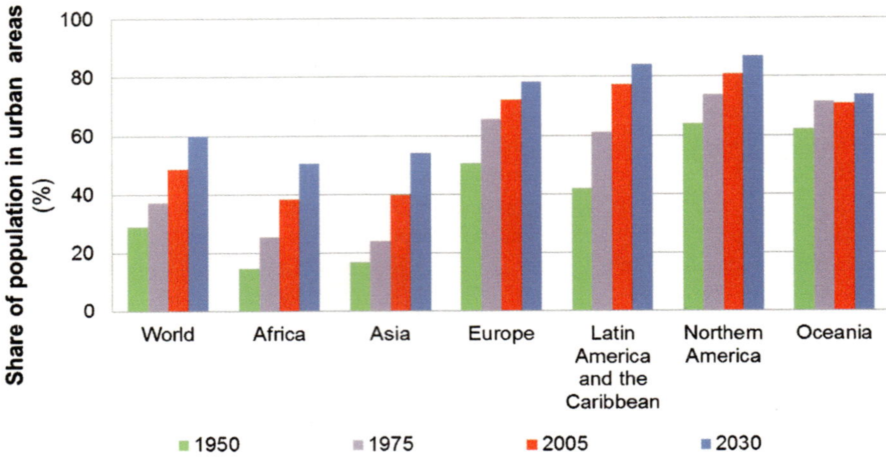

Fig. 2 Urbanisation trends by region (Source: UN DESA 2007)

Urbanisation and the growth of cities are the result of a number of processes, including natural population growth, rural–urban migration and annexation. In general, fertility and natural population growth rates are slightly lower in urban than in rural areas. Thus, the main drivers of urbanisation are rural–urban migration, the expansion of urban areas through annexation, and the transformation and reclassification of rural villages as small urban settlements. The relative importance of these causes of urbanisation and suburbanisation varies both within and between regions and countries (Cohen 2006).

In addition to the general trend of increasing urbanisation, 'mega-cities' are increasing both in number and size (see Fig. 1, right panel). In 1990, only ten cities had populations of over ten million. By 2010, that number had more than doubled, most of them the developing world. However, their growth has slowed recently, reflecting declining population growth rates. In fact, for the foreseeable future, most urban population growth will take place in smaller cities and towns. Large cities will play a significant role in absorbing future population growth, but most urban residents will live in smaller settlements with fewer than 500,000 inhabitants (Kitio 2007).

Recent projections indicate that the most rapid urbanisation can be expected in South Asia, where current levels of urbanisation are fairly low. A study of urbanisation in India, for example, used three scenarios to assess the degree of urbanisation by the end of the century (see Fig. 3). Starting from a level of about 30% in 2010, the projections show quite large variations, ranging from less than 40% under the 'slow' scenario, to over 80% of the population living in urban areas under the 'fast' scenario (O'Neill et al. 2012). Much of this uncertainty stems from assumptions regarding the degree of net in-migration from rural to urban areas. The causes of migration, whether they are 'push' factors that result in people moving out of rural areas, or 'pull' factors that attract them to cities, are relatively little understood.

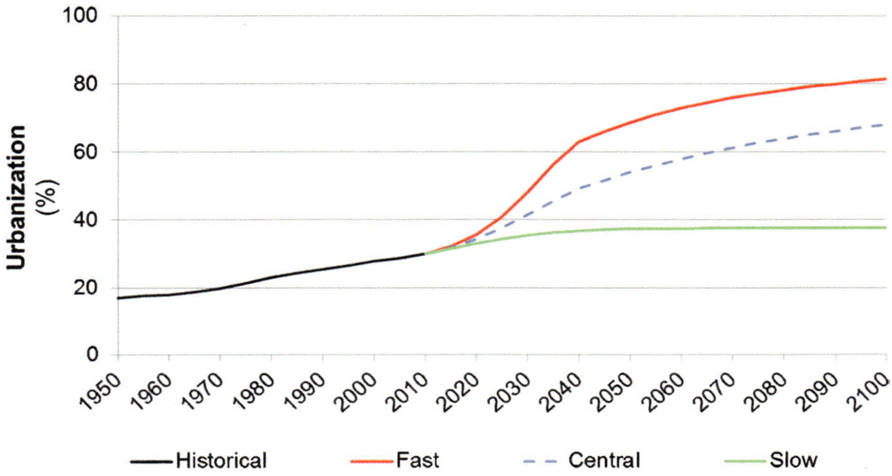

Fig. 3 Urbanisation trends in India: three scenarios (Source: O'Neill et al. 2012)

Thus, there remains great uncertainty concerning how much urban migration is likely to occur in the future. What can be said with more certainty is that India's urban population will increase significantly in the coming decades.

3 Urban Household Energy Access and Use Patterns

A considerable proportion (30%) of the urban populations in low- and middle-income nations lack access to modern fuels for cooking and heating. In the least developed nations, a higher proportion of the urban population has access to electricity, but about half still lack access to modern fuels (UNDP/WHO 2009) (see Table 1). These regional estimates, of course, mask considerable variations between and within countries. Low-income urban households in the least developed countries are less likely to have access to modern forms of energy than richer households (Fall et al. 2008; Dhingra et al. 2008; Ghanadan 2004).

The evidence on urban energy use in developing nations in Africa and Asia suggest that it is common for poor urban households to use a mix of fuels for different end uses (or even for cooking different types of food), and to switch fuels when fuel prices or household incomes change (Pachauri and Jiang 2008; Meikle and North 2005). Regional and seasonal differences in energy use patterns can thus be significant. The available data indicate that some low-income urban households use little or no fossil fuels and electricity (i.e. they use wood, dung, straw and charcoal), while others use kerosene, coal or charcoal, and yet others use gas (bottled or piped) and electricity.

Table 1 Access to electricity and modern fuels in developing countries: share of population with access (%) (Source: UNDP and WHO 2009)

		Rural	Urban
Electricity	All developing countries	59	90
	Least developed countries	13	44
	Sub-Saharan African nations	11	54
Modern fuels	All developing countries	19	70
	Least developed countries	3	27
	Sub-Saharan African nations	5	42

In most middle-income and some low-income nations, most urban residents have access to electricity. By 2002, there was near-universal access to power in most Latin American cities (WEC 2006). Household survey data from India for the year 2008–2009 show that 96% of urban households use electricity (NSSO 2010). In China, survey data for 2001 show that 96% of households use electricity (Pachauri and Jiang 2008). Data from many other nations, including Colombia, Dominican Republic, Egypt, Indonesia, Jordan, Pakistan, and Ukraine, indicate that more than 98% of their urban populations have access to electricity (UNDP and WHO 2009).

Thus, there are many nations where a high proportion of poor urban households have access to electricity. For instance, a study of energy use in slums in Bangkok and Khon Kaen in Thailand found almost 100% had electricity connections (Shrestha et al. 2008). In Mexico in 2000, 91–97% of the poorest (lowest quartile) households in cities along the US border had access to electricity (Peña 2005). In Pakistan, over 80% of the poorest urban households had electricity (ESMAP 2006).

In low- and middle-income countries, poor access to modern fuels and cooking devices in urban regions is more concerning. In the least-developed countries, over two-thirds of urban households still rely on solid fuels such as wood, charcoal and animal residues for cooking and heating. In sub-Saharan Africa, more than half of urban households rely on such fuels (UNDP/WHO 2009). Within these countries, however, there are very large variations in the patterns of energy use among different income groups and socio-economic classes. Thus, as mentioned above, low-income urban households in poor countries and regions still have extremely limited access to modern forms of energy.

4 Energy Access and Use in Rural and Urban Households in China and India

In China and India the patterns of energy access and use among rural and urban households differ significantly. In India, for instance, most rural households still rely on traditional biomass (wood, crop and animal residues) for more than 90% of their final energy needs (see Fig. 4), and use primitive cooking devices that are inefficient and highly polluting. In China, rural households rely on biomass for over 70–75% of their final energy use (Pachauri and Jiang 2008). For urban households in these two countries, however, the patterns of energy use are rather different.

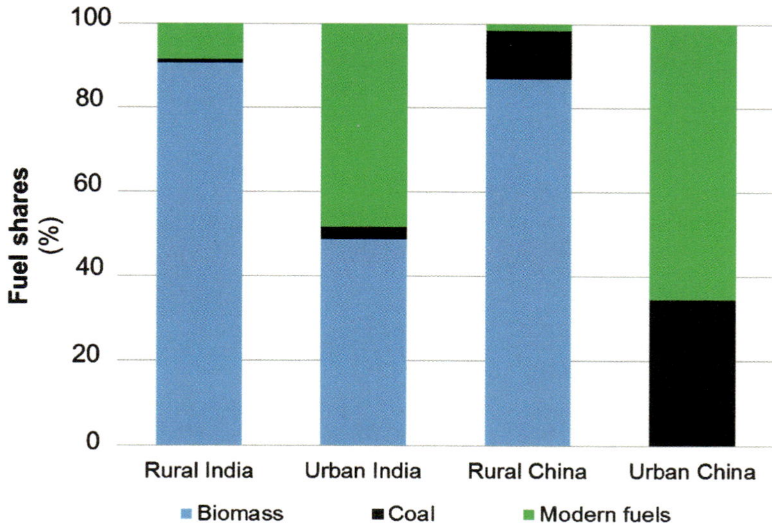

Fig. 4 Household fuel use in rural and urban areas in India and China, 2008 (Source: Modified from Pachauri and Jiang 2008)

In urban India, while households still use biomass to some extent, they use a much higher proportion of modern fuels such as kerosene and gas. In urban China, on the other hand, households no longer use biomass, but coal is still widely used.

It is also useful to compare household energy access and use in different types of urban areas (see Fig. 5; Pachauri and Jiang 2008). In general, the shift away from solid fuels, biomass and coal, towards liquid, gaseous and grid fuels has been more rapid in large metropolitan areas than in small towns. Accordingly, households in large cities obtain a larger share of their total energy needs from modern energy carriers.

Figure 6 shows the changes in household energy use in India and China over the last 25 years. The graphs present the percentage shares of different types of energy in the total final energy used by households. For rural households, there has been very little change, and biomass continues to dominate the mix of fuels used for cooking and heating. But for urban households the story is very different. In China there has been a shift away from coal towards electricity and gas, and in India away from biomass and coal towards gas/oil and electricity. Thus, the transition in household energy use patterns has been quite dramatic in urban areas, whereas in rural areas there has been very little change.

Figure 7 shows the patterns of energy use among households belonging to different income groups in rural and urban areas in India for the year 2004–2005. The mean annual expenditures for each of these income groups are shown on the horizontal axis. Figure 7 reveals that while higher-income groups use more electricity, liquefied petroleum gas (LPG) and other modern fuels, the dependence on biomass remains fairly high among rural households. So even rural households in the top

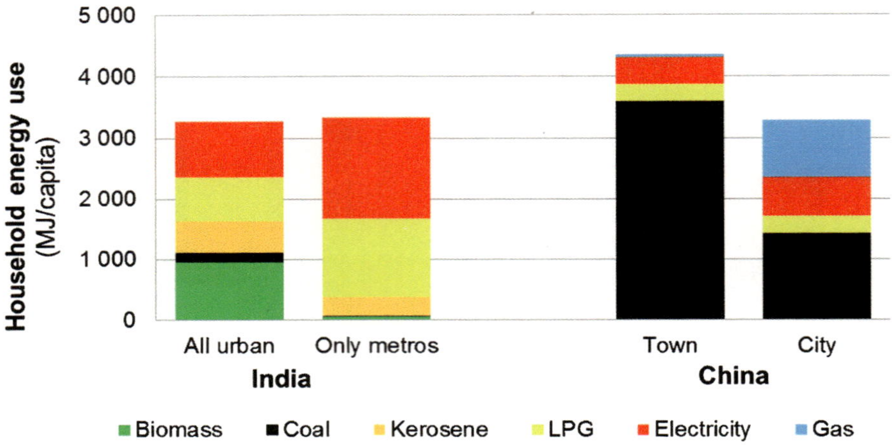

Fig. 5 Variations in urban household energy use, by type of urban area (Source: Modified from Pachauri and Jiang 2008; Note: *MJ* megajoules (10^6 J), *LPG* liquefied petroleum gas)

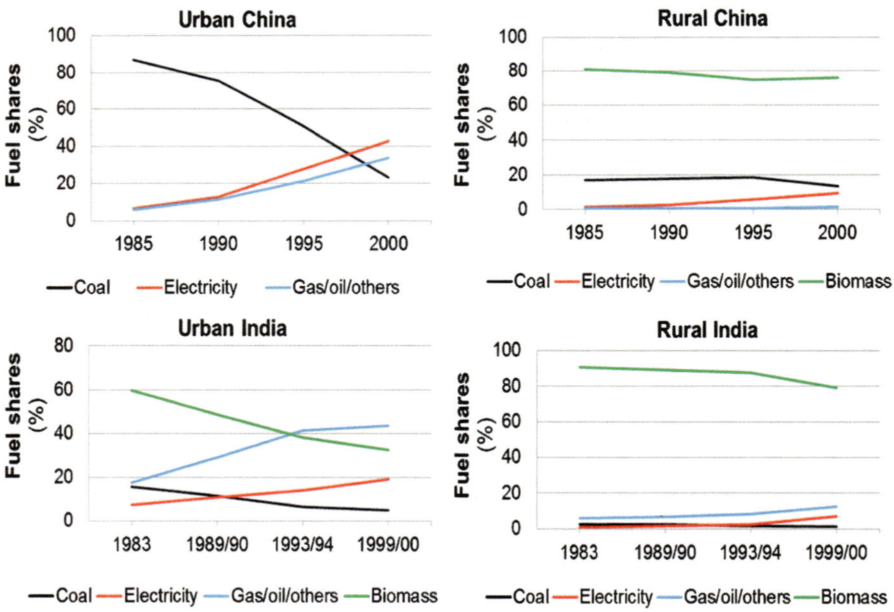

Fig. 6 Household energy use in urban and rural areas in India and China (Source: Modified from Pachauri and Jiang 2008)

income decile still use biomass to meet over 70% of their total energy needs. In urban households, in contrast, there has been a big shift in the pattern of energy use. Whereas the poorest households use biomass to meet over 70% of their energy needs, the richest households use largely LPG and electricity.

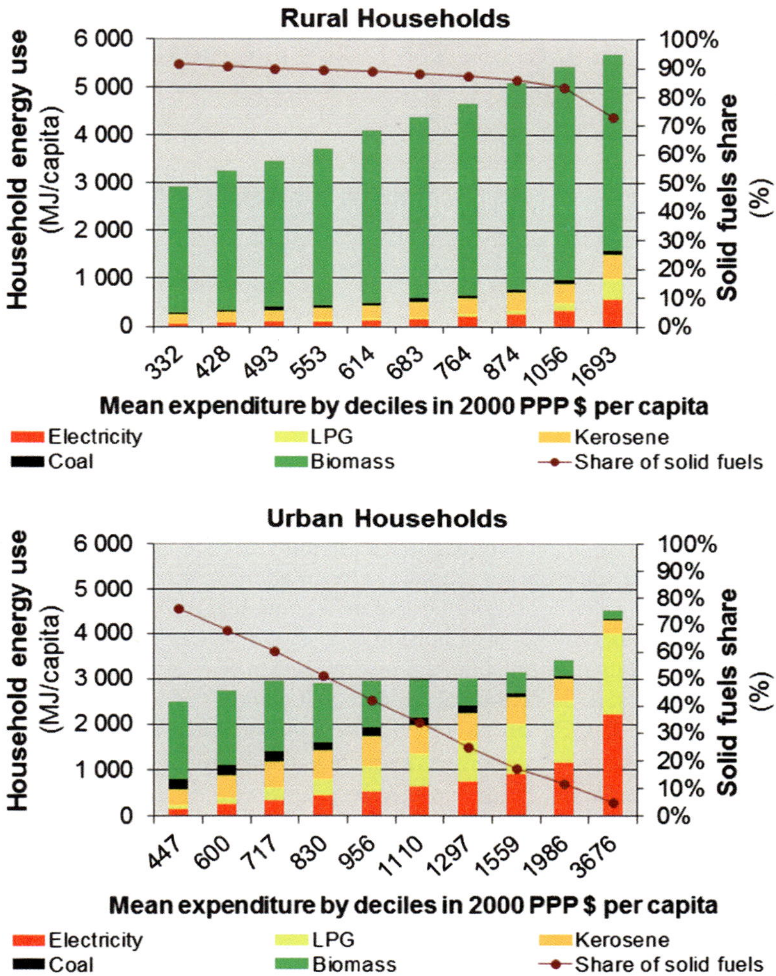

Fig. 7 Household energy use patterns in India, 2004–2005 (Source: Modified from Pachauri and Jiang 2008; Note: *PPP* purchasing power parity, *LPG* liquefied petroleum gas)

There has been a big shift in the pattern of energy consumption in urban households, whereas the same shift is not observed in rural households. In other words, there has been basically little transition in energy use in rural areas. To a large extent, this can be explained by the fact that rural homes lack access to modern forms of energy. Examining the patterns of energy consumption for rural and urban households with similar incomes (e.g. rural decile 5 and urban decile 2 in Fig. 7), it is clear that urban households use much more fossil fuels and electricity than rural households, since they have easier access to them. It is also interesting to note that, on average, rural households use more final energy than urban ones. This is because rural households use more biomass, which they burn in highly inefficient devices, so that large amounts are needed to satisfy low levels of demand. Of course, on average, urban households use more fossil fuels and electricity than rural households.

5 Future Energy Access Scenarios for South Asia

The Global Energy Assessment presents residential energy access scenarios for regions where the lack of access is most acute. The scenarios were constructed using the MESSAGE-ACCESS model (Ekholm et al. 2010; Pachauri et al. forthcoming; Riahi et al. forthcoming), which is a least-cost optimisation model for the residential sector. The model distinguishes between the energy use patterns among ten or more socio-economic groups, including five urban expenditure quintiles and five rural quintiles. Household fuel choices are based on costs, including the price of the energy itself, the fuels available, and the cost of appliances. Since consumers' rate of time preference or discounting horizon varies tremendously according to their income, income-dependent implicit discount rates are used to annualise the costs of appliances for each household group. In addition, in the model, choice is also constrained by the household budget or income level. Since, particularly for rural households, income alone is not the only factor that affects fuel choice, the model includes another novel factor, an 'inconvenience cost' associated with the use of low-quality fuels, as an additional variable that determines fuel choice.

The MESSAGE-ACCESS model was used to construct various scenarios of future energy access and use among different household groups up to 2030. Figure 8 shows some of the results of these projections for South Asia. It can be seen that

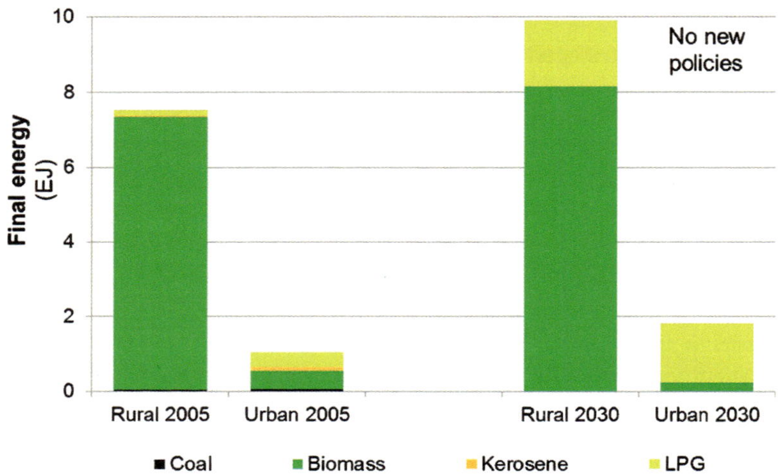

Fig. 8 Rural and urban household fuel use in South Asia, 2005 and 2030 (Source: GEA forthcoming; Note: *EJ* exajoules (10^{18} J), *LPG* liquefied petroleum gas)

even in the 'no new policies' scenario, between 2005 and 2030 the use of biomass in urban areas drops significantly as a result of socio-economic changes, in particular income growth. In rural areas, by contrast, biomass use increases by about 11% during this period, in the case where no new policies are implemented to improve their access to modern forms of energy.

6 Policy Conclusions

The analyses and the scenarios presented in this chapter point to the importance of future patterns of migration, urbanisation and changes in urban density for both the household energy mix and the evolution of energy demand in the coming decades. Both the amounts and types of energy that people use will continue to differ significantly between cities and rural areas. In both developing and least developed countries, projections strongly indicate that urban slums will continue to grow. These areas urgently require investments for provision of energy services. Given the large differences in energy use patterns even within urban areas, it is critical not to lose sight of low-income households who tend to be excluded from the services that most urban residents enjoy. In particular, urban slum populations will need special attention. Concerted efforts and investments will be needed to improve the living conditions of slum populations and their access to more efficient and safer energy carriers.

In the future, urbanisation trends are likely to be influenced by how cities evolve. Rural populations are attracted to cities because they offer better employment opportunities, but better infrastructure and access to services could also play an important role in accelerating migration from rural to urban areas. Without adequate access to clean energy sources in cities, the 'pull' to cities might be weakened, and this could have implications for the rate at which the demographic transition occurs (as there is clear evidence that people in rural areas have larger families than city dwellers).

The lack of access to modern energy has several implications for urban households, including poor air quality, adverse health effects and, as a consequence, high mortality and morbidity rates. These might also influence people's decisions to move to urban areas in the future. Further studies to understand the motivations for migration, and how access to energy and other infrastructure and services might influence such decisions, could provide the policy community with clearer evidence of the importance of improving access to modern energy in both rural and urban areas.

References

Allen, E. L., & Edmonds, J. A. (1981). Energy demand and population changes. *Atlantic Economic Journal, 9*(3), 10–19.

Baker, J. L. (2008). *Urban poverty: A global view* (Urban Paper UP-5). Washington, DC: World Bank. http://go.worldbank.org/2UQ8R8X0H0

Birdsall, N. (1992). *Another look at population and global warming* (Policy research working paper no.1020). Washington, DC: World Bank.
Brew-Hammond, A. (2012). Energy: The missing Millennium Development Goal. In F. L. Toth (Ed.), *Energy for development: Resources, technologies, environment* (pp. 35–43). Dordrecht: Springer.
Cohen, B. (2006). Urbanization in developing countries: Current trends, future projections, and key challenges for sustainability. *Technology in Society, 28*(1–2), 63–80.
Dalton, M., O'Neill, B. C., Prskawetz, A., Jiang, L., & Pitkin, J. (2008). Population aging and future carbon emissions in the United States. *Energy Economics, 30*, 642–675.
Dhingra, C., Gandhi, S., Chaurey, A., & Agarwal, P. K. (2008). Access to clean energy services for the urban and peri-urban poor: A case-study of Delhi, India. *Energy for Sustainable Development, 12*(4), 49–55.
Ekholm, T., Krey, V., Pachauri, S., & Riahi, K. (2010). Determinants of household energy consumption in India. *Energy Policy, 38*, 5696–5707.
ESMAP. (2006). *Pakistan: Household use of commercial energy* (Report No. 320/06). M. Kojima (Ed.). Washington, DC: Energy Sector Management and Assistance Programme (ESMAP)/ World Bank.
Fall, A., Sarr, S., Dafrallah, T., & Ndour, A. (2008). Modern energy access in peri-urban areas of West Africa: The case of Dakar, Senegal. *Energy for Sustainable Development, 12*(4), 22–37.
Ferré, C., Ferreira, F. H. G., & Lanjouw, P. (2010). *Is there a metropolitan bias? the inverse relationship between poverty and city size in selected developing countries* (Policy research working paper 5508). Washington, DC: World Bank.
GEA (Global Energy Assessment). (forthcoming). *The global energy assessment: Toward a more sustainable future.* Laxenburg/Cambridge: IIASA/Cambridge University Press.
Ghanadan, R. (2004). *Negotiating reforms at home: Natural resources and the politics of energy access in Urban Tanzania* (Working Paper). Berkeley: Energy and Resources Group, University of California. www.cnr.berkeley.edu/~alyons/breslauer/papers/ghanadan_politics-energy-tanzania.pdf
Gruebler, A., Bai, X., Buettner, T., Dhakal, S., Fisk, D. J., et al. (forthcoming). Urban energy systems. In *The global energy assessment: Toward a more sustainable future.* Laxenburg/ Cambridge: IIASA/Cambridge University Press.
Ha, M., Williams, S. E., & Webber, M. J. (1993). Utility averaging policies: Impact on consumer's energy usage. *Journal of Consumer Affairs, 27*(2), 284–299.
IEA (International Energy Agency). (2009). *CO2 Emissions from fuel combustion: Highlights* (IEA statistics series). Paris: OECD/IEA.
IEA (International Energy Agency). (2010). *World energy outlook 2010.* Paris: OECD/IEA.
Kitio, V. (2007, November 21–23). *Energy efficiency and poverty reduction in urban areas.* Presentation at the global forum on sustainable energy: Energy efficiency for developing countries, Vienna.www.gfse.at/fileadmin/dam/gfse/gfse7/Plenary6/1_GFSE7_Presentation_UN-Habitat_Kitio.pdf
Lenzen, M., Dey, C., & Foran, B. (2004). Energy requirements of Sydney households. *Ecological Economics, 49*(3), 375–399.
Lenzen, M., Wier, M., Cohen, C., Hayami, H., Pachauri, S., & Schaeffer, R. (2006). A comparative multivariate analysis of household energy requirements in Australia, Brazil, Denmark, India and Japan. *Energy, 31*(2–3), 181–207.
Luna, N., & Gomelsky, R. (2012). Development, infrastructure and energy: Exploring the linkages in Latin America. In F. L. Toth (Ed.), *Energy for development: Resources, technologies, environment* (pp. 95–112). Dordrecht: Springer.
Martin, G., & Richels, R. (1979). Assessing energy policy models: Current state and future directions. *Annual Review of Energy, 4*, 467–500.
Meikle, S., & North, P. (2005). *A study of the impact of energy use on poor women and girls' livelihoods in Arusha, Tanzania* (R8321). London: Department for International Development (DFID).
Nakicenovic, N., & Riahi, K. (2001). An assessment of technological change across selected energy scenarios. In *Energy technologies for the twenty-first century.* London: World Energy Council.

Nakicenovic, N., & Riahi, K. (2003). *Model runs with MESSAGE in the context of the further development of the Kyoto Protocol.* Final Report to the German Advisory Council on Global Change (WGBU), Berlin, Germany.

Nakicenovic, N., Kolp, P., Riahi, K., Kainuma, M., & Hanaoka, T. (2006). Assessment of emissions scenarios revisited. *Environmental Economics and Policy Studies, 7*(3), 137–173.

NSSO (National Sample Survey Office). (2010). *Housing condition and amenities in India 2008–09* (NSS Report No. 535 (65/1.2/1)). New Delhi: NSSO/Ministry of Statistics and Programme Implementation/Government of India.

O'Neill, B. C. (2003). Climate change and population: The future. In P. Demeny & G. McNicoll (Eds.), *Encyclopedia of population* (rev. ed., vol. 1, pp.149–152). Woodbridge: Macmillan Reference USA.

O'Neill, B. C. (2005). US socio-economic futures. In F. Müller & A. Riechel (Eds.), *Options for future climate policy: Transatlantic perspectives* (pp. 46–55). Berlin: Stiftung Wissenschaft und Politik.

O'Neill, B. C., & Chen, B. (2002). Demographic determinants of household energy use in the United States. In Methods of population–environment analysis, supplement to *Population and Development Review,* 28, 53–88.

O'Neill, B. C., Dalton, M., Fuchs, R., Jiang, L., Pachauri, S., & Zigova, K. (2010). Global demographic trends and future carbon emissions. *Proceedings of the National Academy of Sciences, 107*, 17521–17526.

O'Neill, B. C., Ren, X., Jiang, L., & Dalton, M. (2012). The effect of urbanization on energy use in India and China in the iPETS model. *Energy Economics* (accepted).

OECD (Organisation for Economic Co-operation and Development). (2001). *Environmental outlook.* Paris: OECD.

OECD (Organisation for Economic Co-operation and Development). (2007). *Environmental performance reviews: Australia.* Paris: OECD.

Pachauri, S. (2004). An analysis of cross-sectional variations in total household energy requirements in India using micro survey data. *Energy Policy, 32*(15), 1723–1735.

Pachauri, S., & Jiang, L. (2008). The household energy transition in India and China. *Energy Policy, 36*(11), 4022–4035.

Pachauri, S., Brew-Hammond, A., Barnes, D. F., Bouille, D. H., Gitonga, S., et al. (forthcoming). Energy access for development. In *The global energy assessment: Toward a more sustainable future.* Laxenburg/Cambridge: IIASA/Cambridge University Press.

Peña, S. (2005). Recent developments in urban marginality along Mexico's northern border. *Habit International, 29*(2), 285–301.

Price, L., de la Rue du Can, S., Sinton, J., Worrell, E., Nan, Z., et al. (2006). *Sectoral trends in global energy use and greenhouse gas emissions.* Berkeley: Lawrence Berkeley National Laboratory.

Prinn, R. G., Jacoby, H. D., Sokolov, A. P., Wang, C., Xiao, X., et al. (1999). Integrated global system model (IGSM) for climate policy assessment: Feedbacks and sensitivity studies. *Climate Change, 41*(3), 469–549.

Prskawetz, A., Jiang, L., & O'Neill, B. (2004). Demographic composition and projections of car use in Austria. In *Vienna yearbook of population research*, pp. 1–21. www.demogr.mpg.de/Papers/workshops/010518_paper07.pdf. Accessed 12 May 2011.

Ravallion, M., Chen, S., & Sangraula, P. (2007). *New evidence on the urbanization of global poverty* (Policy research working paper 4199). Washington, DC: World Bank.

Reilly, J. M., Edmonds, J. A., Gardner, R. H., & Brenkert, A. L. (1987). Uncertainty analysis of the AEA/ORAU CO_2 emissions model. *Energy Journal, 8*(3), 1–29.

Reilly, J., Prinn, R. G., Harnish, J., Fitzmaurice, J., Jacoby, H. D, et al. (1999). *Multi-gas assessment of the Kyoto Protocol* (Report No. 45), MIT Joint Program on the Science and Policy of Global Change. Cambridge: Massachusetts Institute of Technology.

Riahi, K., Gruebler, A., & Nakicenovic, N. (2007). Scenarios of long-term socio-economic and environmental development under climate stabilization. In K. Riahi & N. Nakicenovic (Eds.) Greenhouse gases: An integrated assessment, special issue of *Technological Forecasting and Social Change, 74*(7), 887–935.

Riahi, K., Dentener, F., Gielen, D., Grubler, A., Jewell, J., et al. (forthcoming). Energy pathways for sustainable development. In *The global energy assessment: Toward a more sustainable future*. Laxenburg/Cambridge: IIASA/Cambridge University Press.

Shrestha, R. M., Kumar, S., Martin, S., & Dhakal, A. (2008). Modern energy use by the urban poor in Thailand: A study of slum households in two cities. *Energy for Sustainable Development, 12*(4), 5–13.

Tonn, B., & Eisenberg, J. (2007). The aging US population and residential energy demand. *Energy Policy, 35*(1), 743–745.

UNDP & WHO (United Nations Development Programme & World Health Organization). (2009). *The energy access situation in developing countries: A review focusing on the least developed countries and Sub-Saharan Africa*. New York: UNDP/WHO.

UN, DESA (United Nations Department of Social and Economic Affairs). (2007). *World population prospects: The 2007 revision*. New York: UN.

WEC (World Energy Council). (2003). *Drivers of the energy scene*. London: WEC.

WEC (World Energy Council). (2006). *Alleviating urban energy poverty in Latin America*. London: WEC.

Winistorfer, P., Chen, Z., Lippke, B., & Stevens, N. (2005). Energy consumption and greenhouse gas emissions related to the use, maintenance, and disposal of a residential structure. Consortium for Research on Renewable Industrial Materials (CORRIM), special issue. *Wood and Fiber Science, 37*, 128–139.

World Bank. (2010). *Systems of cities: Harnessing urbanization for growth & poverty alleviation.* Washington, DC: World Bank Urban & Local Government Strategy. http://go.worldbank.org/JXKR1EI2Q0

Development, Infrastructure and Energy: Exploring the Linkages in Latin America

Nestor Luna and Roberto Gomelsky

Abstract This chapter explores conceptual and methodological approaches to the linkages between economic development, infrastructure and energy in Latin America and the Caribbean. It demonstrates a high correlation between energy consumption and GDP per capita, indicating the effect of economic growth on energy demand. This linkage materialises through infrastructure and equipment technologies, operational characteristics and actual utilisation. The energy infrastructure is also shaped by the distribution of population and the concentration of GDP. Power transmission and distribution grids, natural gas pipelines and other infrastructure are oriented to deliver energy products to these major markets. The chapter proposes a new methodological approach to energy planning and modelling that can be linked to infrastructure planning for the transportation, water, industry, communications and other sectors.

1 Introduction

The energy sector provides a variety of goods and services demanded by the broader economy. The sector is a critical ingredient in the production of all the goods and services required by society, as well as satisfying the energy needs of individuals, households and communities. As a consequence, the pattern of socio-economic

N. Luna
Latin American Energy Organization (OLADE), Av. Occidental, Edificio OLADE, Quito, Ecuador
e-mail: nestor.luna@olade.org

R. Gomelsky (✉)
Freelancer Consultant, Pasaje Lombeyda # 7, San Rafael Quito, Ecuador
e-mail: gomelsky@uio.satnet.net

F.L. Toth (ed.), *Energy for Development: Resources, Technologies, Environment*, Environment & Policy 54, DOI 10.1007/978-94-007-4162-1_8,

development determines the basic energy requirements, including the energy import/ export balance, production, transportation and the distribution infrastructure. When properly planned, the energy sector should be able meet the demand for energy when it is needed, at the lowest possible cost, while minimising negative social and environmental impacts.

Strong linkages between energy and economic growth can lead to overall improvements in the quality of life, by meeting the basic needs of society. Such linkages also lead to the development of adequate infrastructure and equipment for all energy-demanding sectors – transportation, water supply, communications, waste treatment, industry, commerce, etc. – and by final consumers, including buildings and household goods (see Calderón and Servén 2004).

Each energy-demanding sector develops its own infrastructure, which requires energy and equipment for its construction and operation. Such infrastructure is highly energy-intensive to build, it is often 'long lived' and expensive to replace, and it can shape the types and forms of energy that are needed. Thus, the type and amount of energy consumed, and the efficiencies and environmental impacts of different consumption processes, will be shaped by the technologies applied in each sector. Analyses of these linkages are crucial because the nature of the linkages and their quantification will determine the capital requirements and operational costs of energy exploration, exploitation, storage, transformation and distribution.

The energy sector is highly capital intensive, and energy projects have long development periods. Good investment decisions require careful, thorough and permanent planning processes. This also means that the major demand sectors need to be carefully planned to facilitate the formulation and implementation of energy plans. Unfortunately, this so-called 'planning linkage' is largely missing in most countries in Latin America and the Caribbean (LAC).

This chapter addresses the conceptual and methodological aspects of the linkages between economic development, infrastructure and energy in the LAC region. Section 2 discusses some quantitative indicators of energy–economy linkages in the region, and Sect. 3 addresses some methodological approaches to economic development and infrastructure. Section 4 analyses the relationships between energy and infrastructure, and the critical need for data and information. It also provides examples of energy demand modelling based on the drivers for major consumption sectors, and of quantitative analyses of energy infrastructure. Finally, Sect. 5 presents a summary and conclusions.

2 Energy–Economy Linkages in the LAC Region

2.1 Economic and Social Indicators

The main lesson emerging from two decades of observations in the LAC region is that all governments need to formulate policies that synchronise private and public interests. They also suggest that all countries need to monitor the advances of businesses, and to develop operational investment plans to ensure secure energy supplies over the short and medium term, within the context of sustainable development.

Over the long term, scenarios can be used to examine ways in which energy demand growth can be managed through feasible actions that will encourage rational energy consumption in all sectors. Such actions could include measures to improve energy-use efficiency, the introduction of new technologies and substitutions between energy sources, taking into account the local resources available and the energy production and transformation infrastructure.

The future energy and economic scenarios for the region are based on the study by Gomelsky et al. (2009) prepared for the Latin American Energy Organization (OLADE), with the support of the World Bank. This five-volume study includes two macroeconomic scenarios and corresponding integrated energy demand and supply scenarios for each of the 26 member countries of OLADE, as well as analyses at the subregional level and for the region as a whole.

Despite the steady economic growth in the LAC region in recent years, there are still significant differences between countries in gross domestic product (GDP) per capita originating in the availability of natural resources and the level of integration in the agricultural, industrial and mining sectors. Many LAC countries are exporters of primary products and importers of manufactured and capital goods, and have limited access to new technologies due to the lack of a qualified work force.

Under OLADE's rapid economic growth scenario (Gomelsky et al. 2009), GDP per capita will increase, which will lead to changes in energy consumption in the major consuming sectors, and in the energy production, transformation and transportation infrastructures. Looking at the high-growth scenarios of individual countries over the long term, the gap in GDP per capita will significantly narrow, but important differences between countries will remain. Nevertheless, higher incomes and better living standards do not necessarily lead to increased energy demand, because of the introduction of new, more efficient technologies and enhanced operational practices. In other words, as economies grow, it is essential to reduce the energy demand–GDP elasticity.

2.2 *Energy and Economy*

Over the last four decades, energy consumption has grown steadily at the regional level, but especially in the biggest economies, including Brazil, Mexico and Argentina, which together represent more than half of total consumption (see Fig. 1). They also have the most diversified energy matrices and the highest levels of technology development, e.g. in oil exploration, nuclear energy and natural gas. The subregions in Fig. 1 correspond to those employed in the OLADE Information System (SIEE) (OLADE 2010):

- Andean zone: Chile, Bolivia, Peru, Ecuador, Colombia and Venezuela;
- Caribbean: Cuba, Haiti, Dominican Republic, Barbados, Jamaica, Trinidad & Tobago, Grenada, Suriname and Guyana;

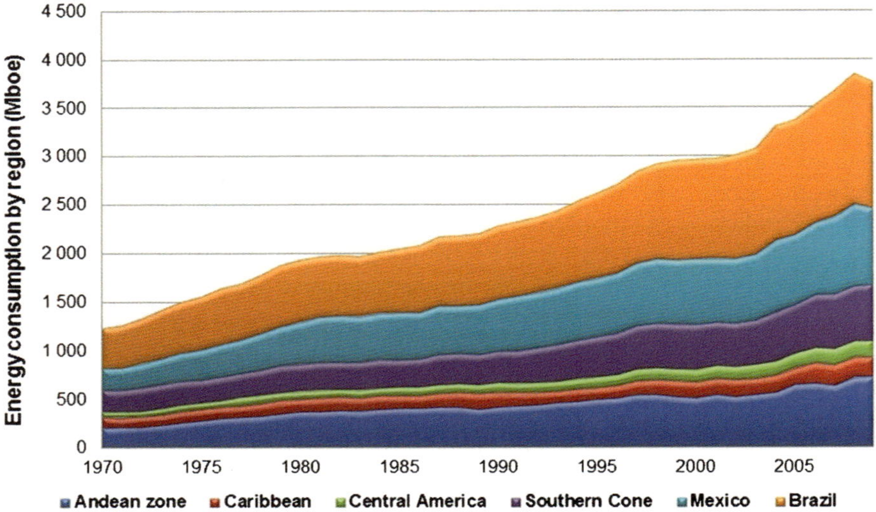

Fig. 1 Energy consumption in Latin America and the Caribbean, by region, 1970–2006 (Data source: OLADE 2010; Note: *Mboe* million barrels of oil equivalent)

- Central America: Belize, Costa Rica, El Salvador, Guatemala, Honduras, Nicaragua and Panama; and
- Southern Cone: Argentina, Chile, Paraguay and Uruguay.

Due to the moderate rate of GDP growth and the lack of clear and consistent policies on energy efficiency in most LAC countries, the average energy intensity of the region has improved only slightly as a result of the natural evolution of available technologies for infrastructure and equipment and their application to some final energy uses in different sectors.

On the energy–economy side, there is a strong linkage between energy consumption per capita and GDP per capita. The latter is one of the components of the Human Development Index and is a good indicator of average income growth. This does not necessarily reflect improvements in living conditions, but is an indicator of energy demand at the final consumption level. Considering the energy consumption per capita and the average GDP per capita across the region between 1990 and 2008, a very high statistical correlation can be found, indicating that economic growth is an important driver of energy demand.

Another important variable is capital stock, a macroeconomic aggregate that reflects the investment in fixed capital made by the economy as a whole. Consequently, this is an indicator of infrastructure and equipment development that is a good representation of energy infrastructure–economy linkages. The trend in the LAC region over the period 1990–2008 shows a very good correlation along a logarithmic function, indicating a declining growth rate of energy consumption

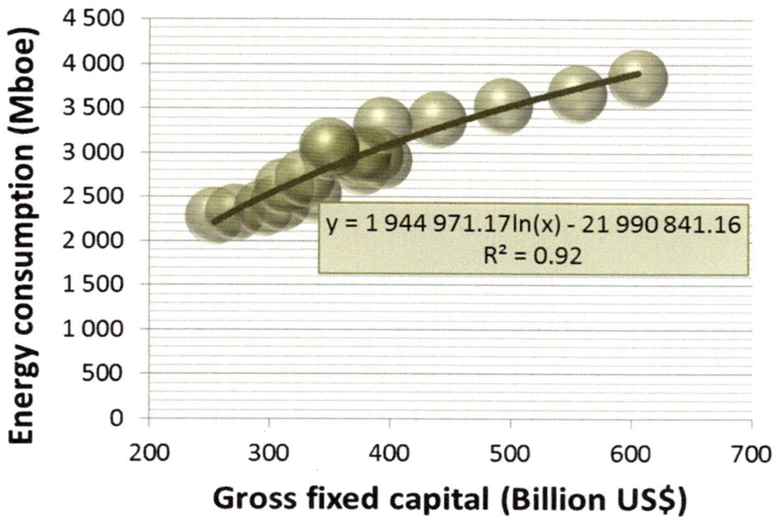

Fig. 2 Energy consumption and gross capital stock in the LAC region, 1990–2008 (Data sources: OLADE 2010; ECLAC 2010; Note: Each point corresponds to one year; *Mboe* million barrels of oil equivalent)

with increasing capital stock (Fig. 2), following the steady decline in energy intensity indicated above.

Looking at probable medium- to long-term developments, emphasis must be placed on improving the overall energy efficiency of the economy, following the high-growth scenario, and on 'disconnecting' energy demand growth from economic growth. It will also be necessary to encourage structural changes in demand through changes in infrastructure and equipment that will lead to a lower energy demand–GDP elasticity. Over the period 2007–2032, the energy–GDP elasticity for the LAC region may be reduced by 15% under an energy efficiency scenario with structural changes, compared with a 'business-as-usual' scenario. With the exception of the Caribbean, in some areas of Mexico, Central and South America, the reduction may be as high as 20–30%. These changes will also be reflected in the energy intensity of the economy, which may decline almost three times as fast in the energy efficiency scenario than in the business as usual scenario.

Regarding the composition of total consumption, even in the energy efficiency with substitution scenario, the main sources of energy will continue to be oil and gas, which will account for 62% by 2032, with some substitution of oil products by natural gas. The differences between some parts of the region (see Table 1 and Fig. 3) reflect the continuing importance of transportation and industrial infrastructure and equipment on the demand side, and of the oil and gas industry and power generation on the supply side.

Table 1 Latin America & Caribbean: energy demand projections by region (Source: Gomelsky et al. 2009)

	Mexico		Central America		South America		Caribbean		LAC region	
	2007	2032	2007	2032	2007	2032	2007	2032	2007	2032
Solid fuels	257	513	19	91	759	1,788	19	60	1,054	2,452
Other renewables	0	186	1	81	0	376	0	40	1	683
Other sources			44	19	261	505	7	17	312	540
Oil products	2,984	3,814	425	829	7,612	15,028	736	1,884	11,757	21,555
Natural gas	1,157	2,562			2,638	6,612	604	1,889	4,399	11,063
Electricity	711	1,493	175	328	2,762	7,241	165	514	3,812	9,576
Biomass	248	143	319	314	2,654	4,793	167	260	3,387	5,510
Biofuels	0	138	126	13	377	847	0	29	503	1,027
TOTAL	**5,356**	**8,849**	**1,109**	**1,674**	**17,062**	**37,190**	**1,698**	**4,693**	**25,226**	**52,406**

Note: *PJ* petajoule (1 PJ = 10^{15} J)

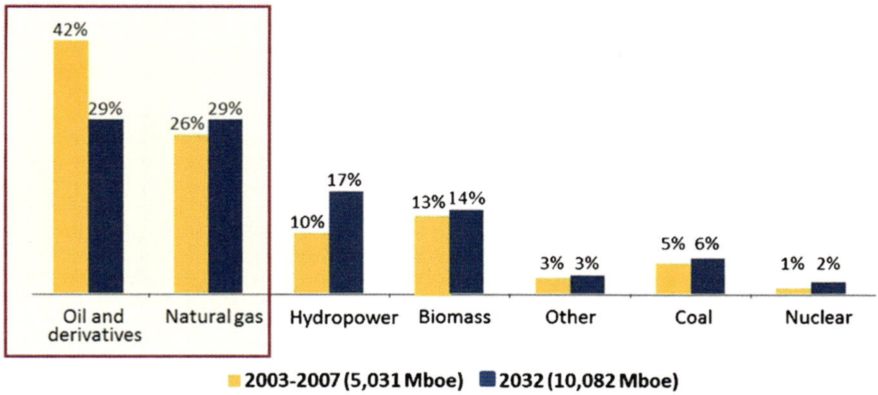

Fig. 3 Projected shares of fuels in the energy balance of the LAC region in 2032 (Data sources: OLADE 2010; Gomelsky et al. 2009; Note: *Mboe* million barrels of oil equivalent)

3 Economic Development, Infrastructure and Energy

3.1 The Integrated Energy–Development Approach

In any given country or region, the energy system represents the flows of energy from natural resources to the emissions and residues that energy demand and supply produce (see Fig. 4). The energy flows shown in the figure represent the energy balance, but that is only a part of the energy matrix or energy structure of a country or region. By widening the scope of components of the energy matrix, it is possible to obtain an integrated vision of all the energy segments that have to be considered in long-term projections for each pattern or scenario analysed, and of the environmental and other impacts to be quantified.

There is a two-way linkage between energy and economic development. On the demand side, economic development requires energy both for building and utilising infrastructure and equipment in different sectors (e.g. transportation, primary and industrial production, water, communications, etc.) as well as for private households. On the supply side, satisfying energy demand requires investments and incurs operational costs all along the energy chain, from the production of primary energy from natural resources, its transformation to secondary energy (such as electricity and oil products) and its transportation and distribution to final consumers.

Considering the flows in the opposite direction, energy investments require financial resources that come from energy tariffs paid by consumers, equity investments by private investors or governments, and loans provided by the financial markets. The international energy trade also has an impact at the macroeconomic level because of energy imports and exports.

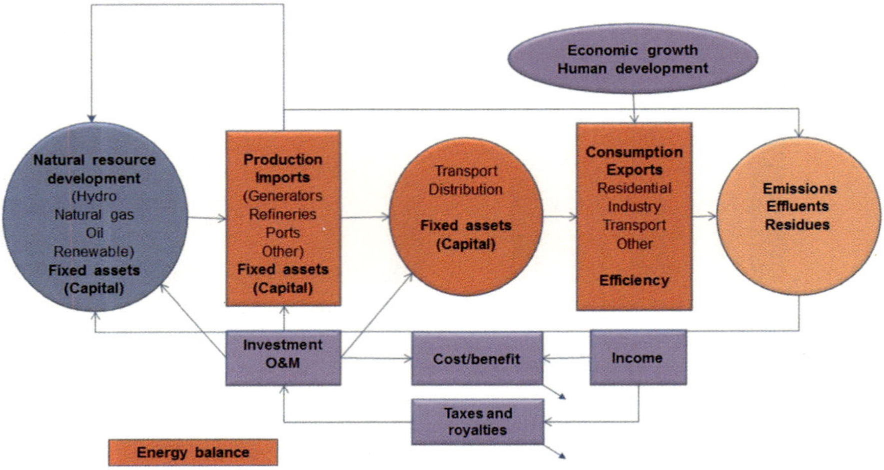

Fig. 4 Energy structure: society, economy and energy linkages (Note: *O&M* Operations and maintenance)

Energy demand is often directly linked to economic variables that act as drivers of final consumption. Infrastructure is not often considered as a direct variable in energy modelling mainly because of the lack of appropriate and reliable data and relevant analyses, and the absence of infrastructure planning. But in the context of socioeconomic development, reflected in economic growth and overall improvements in living standards, the technology that is utilised in infrastructure and equipment will finally determine the energy consumed in different sectors and for different uses. In other words, energy demand is driven by infrastructure and equipment and their utilisation (Fig. 5).

3.2 Energy Planning Requires Infrastructure Planning

To examine economy–technology–infrastructure–energy linkages, a new methodological approach to energy planning and modelling is needed that is in turn linked to infrastructure planning – a kind of integrated infrastructure planning that includes transportation, water, industry, communications and other sectors, as well as energy.

Because of its capital-intensive and long-term nature, the energy sector requires long-range policies, strategies, investment decisions and project scheduling far in advance of project start-up, i.e. a continuous and permanent process of decision-making under uncertain conditions. Thus rational, consistent and continuous planning processes are needed in which long-term energy policies and strategies can be carefully defined. This requires a well-defined regulatory framework and a clear orientation towards economic agents, whether private or state-owned, in order to

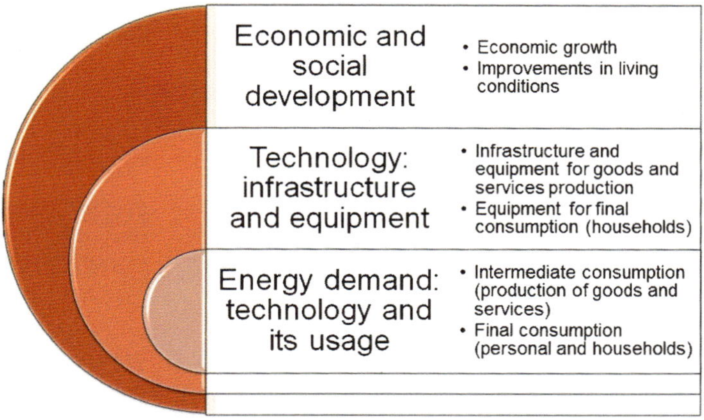

Fig. 5 Infrastructure as an intermediate variable driving energy demand

allow them take the right decisions, under any economic model, institutional organisation, level of regulation, or government control or involvement in economic activities. Many LAC countries, however, have wrongly assumed that the production and operation of the energy sector can be left to private agents, requiring only the regulation of monopolies, and giving total freedom to competitive markets, ignoring the essential role of governments in energy policy and planning.

This weakening of policy-making processes in most LAC countries has resulted in the failure to maintain information systems that are essential for making projections, and for planning analyses that result in good policy decisions and long-term development strategies. Although the situation is changing, and planning systems are gradually being restored, there are several concepts related to energy projections and planning that must be adapted to changing conditions, taking advantage of lessons learnt in the past.

First, it is important to differentiate between projections and planning. A *projection* is a long-term strategic study, usually 25–30 years or more, that uses a number of future scenarios to evaluate the advantages and disadvantages of different energy development strategies. Projections are not intended to be used to decide on investment plans or on the construction of specific projects, but to evaluate the strategic implications of different energy policies. In the case of oil-importing countries, for example, what is the best strategy, to import crude oil and refine it, or to import refined oil products? What detailed studies will be needed at the planning level for exploring the impacts of the substitution of private vehicles by public mass transportation, and for planning the expansion of the infrastructure? Long-term projections can help to answer such questions, and provide orientation for decision makers to enable them to select the best options and then to formulate specific plans.

Planning is a process in which decisions on actual investments in and financing for specific infrastructure projects are made, and to establish a regulatory framework and conditions that will lead to the achievement of quantitative goals over the

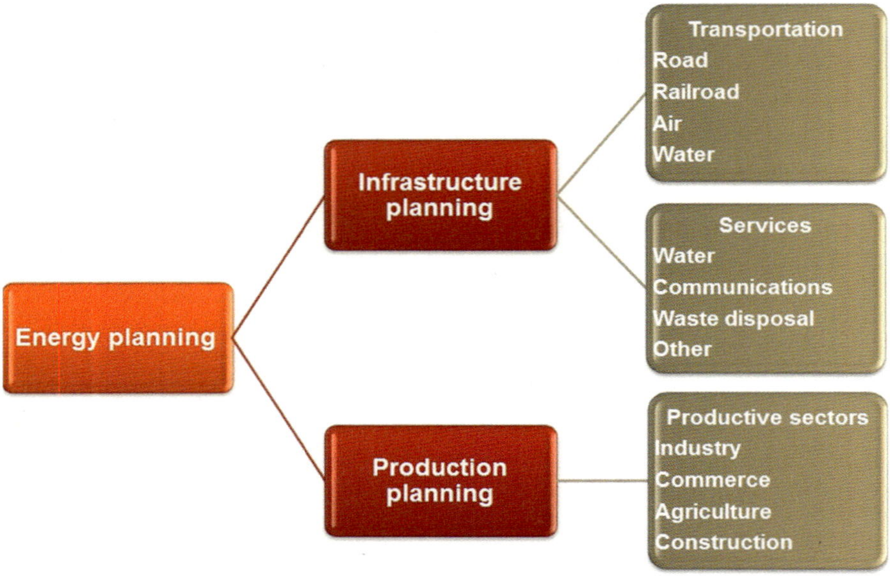

Fig. 6 Integrated energy–infrastructure planning approach

medium and long term. Usually the planning time frame is shorter than those covered in projections, and in most LAC countries is around 10–15 years.

To attain the desired *future* end point, the approach means working from the future back to the present. This means deciding on which long-term objectives are to be achieved, and then going back and setting up ways to achieve them, rather than forecasting or predicting the future. The process requires well-established procedures for monitoring the evolution of the system, and the technical ability and authority to make necessary adjustments at the right times. Within the framework of energy projections and planning that leads to the selection of technologies and investments in energy production, transformation and distribution, it is also important to define plans for the development of infrastructure and production capacity in the major consuming sectors. In other words, energy planning requires infrastructure planning, within an integrated energy–infrastructure planning system (see Fig. 6).

In reality, however, this methodological approach is rarely used in the LAC region. Detailed information on energy consumption by uses and sources is normally lacking, and planning to meet the needs of the major consuming sectors is partial, and restricted to the development of specific projects. Information on existing infrastructure is also often very limited. Many countries carry out studies and undertake regular energy planning activities, but across the LAC region there is little detailed information about the demand side because of the traditional and prevailing supply-side view. As a result, there is limited knowledge of the social, economic and demand profiles of energy-consuming sectors.

On the methodological side, a number of models can be used for integrated analyses of the energy sector in long-range projections or planning studies. Other models can help in the selection of investments in, for instance, power generation and transmission, oil refining, or in evaluations of non-renewable resources like oil and gas. But there are no similar models for planning in the major energy-consuming sectors. Some good examples of integrated energy planning models are the Model for Analysis of Energy Demand (MAED-2) (IAEA 2006), and the Model of Energy Supply Strategy Alternatives and their General Environmental Impacts (MESSAGE) (IAEA 2009), both developed by the International Atomic Energy Agency. Another is the Long-range Energy Analysis Planning System (LEAP), an integrated demand–supply model developed by the Stockholm Environment Institute US Center (Heaps 2008). Although there are models for transportation planning, in most cases they are not fully applied and there have been few comprehensive studies of transportation infrastructure planning, or of the impacts of such developments on future energy demand.

In summary, robust efforts are needed to improve the information, databases and planning methodologies necessary for integrated energy–infrastructure modelling, involving the energy sector and the major consuming sectors.

4 Energy and Infrastructure

Planning and building infrastructure have direct energy implications. Here we consider new methodologies and models that can help meet the challenge of integrated energy–infrastructure planning.

4.1 Energy Demand

4.1.1 Infrastructure and Equipment Drivers

The demand for energy is driven by infrastructure and its technology but, as information is often not available, it is possible to develop a first approach to energy demand modelling using related variables such as the transportation vehicle fleet or economic indicators representing infrastructure and its operation. It is critically important to improve both methodologies and planning tools, and the availability of information and databases. Because of these constraints, some LAC countries approach demand modelling in some major consuming sectors using driving variables for which information is available. Although this is not an ideal solution, it is far better than using aggregate estimates. Energy demand models have been used in a number of national and regional studies, and could be substantially improved as better information and methodologies become available.

In the case of transportation, for example, a good approach to energy demand modelling involves considering the vehicle fleet, making it possible to break

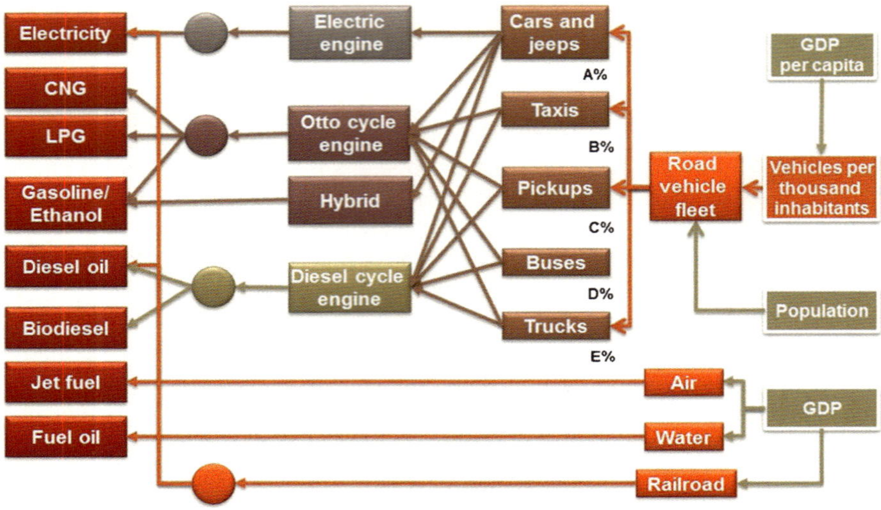

Fig. 7 Example of demand modelling: transportation (Notes: A%, B%, C%, D% and E% are the shares of each type of vehicle in the total fleet; *CNG* compressed natural gas for use in vehicles, *LPG* liquefied petroleum gas, *GDP* gross domestic product)

down energy consumption by type of vehicle, type of engine and different fuels (see Fig. 7).

GDP per capita is a driver that is strongly linked to the vehicle fleet per capita (see below). When the total fleet is broken down by type of vehicle and engine, and then assuming substitutions between energy sources for each type of engine and vehicle, energy consumption by road transportation can be estimated. Air, maritime and railroad consumption can be estimated separately, but as new mass transportation systems are introduced, this should result in lower energy consumption per vehicle used for passengers and freight.

A better way to model the transportation sector is to start with the energy demands of passenger and freight transportation (passenger-km and tonne-km), using an origin and destination matrix for the most important routes, and making intermodal substitutions to make the system more efficient. Again, this requires detailed data and transportation planning systems that consider the energy implications of decisions aimed at improving the efficiency of transportation systems.

The industry sector provides another example. Ideally, the energy required by different branches of industry needs to be determined because different types of industry, such as steel and food processing, have quite different consumption profiles. The next step is to break down energy consumption by uses (lighting, heating, refrigeration, etc.) in order to analyse the efficiency of different technologies and operations in specific industrial processes. But once again, the problem is that this kind of information is often lacking. For less detailed modelling of industrial consumption, one option is to estimate thermal usage of energy and specific electricity uses, and then to break down competing sources of energy for thermal uses (see Fig. 8).

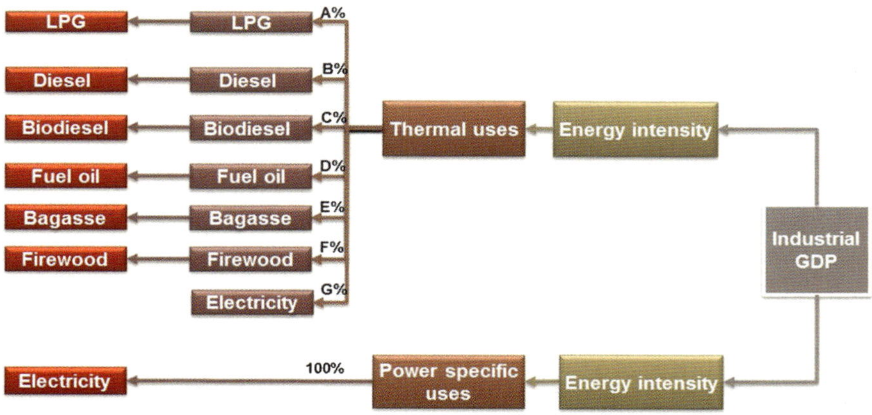

Fig. 8 Example of demand modelling: industry (Notes: A%, B%, C%, D%, E%, F% and G% are the shares of each fuel in thermal uses; *LPG* liquefied petroleum gas, *GDP* gross domestic product)

4.1.2 Energy Efficiency

In many LAC countries, energy consumption in different sectors is generally inefficient, usually because it was not considered important or relevant when equipment was acquired or infrastructure was built. Over the last four decades, there have been many fluctuations in international energy markets, yet the real price of oil has remained more or less the same (see IEA 2009a, b). However, the real price of oil is likely to increase in the future, as rising prices of other raw materials, such as the metals and other materials used to build energy infrastructure, will make any source of energy more expensive over the medium to long term.

According to scenarios developed by the US Department of Energy (EIA 2010), by 2035 the oil price may range from around US$50 per barrel (constant 2008 US dollars) in the low scenario to US$200 per barrel in the high scenario, and around US$130 per barrel in the reference scenario. This confirms the trend towards higher oil prices in real terms observed over the past 20 years, but with some level of attenuation. The most reasonable estimate will probably be between the US DOE reference and high scenarios. On top of the energy price issue, there is strong pressure to improve the efficiency of energy use, arising from environmental concerns. Many countries have introduced policies and strategies to improve energy efficiency, together with stronger regulations requiring all sectors to comply with environmental standards.

The situation is quite different in many Latin American countries. Improved energy efficiency will require regulations and standards that apply to all consuming sectors, not only with regard to the operation of infrastructure and equipment but also its construction. The energy dimension has to be included at all stages, beginning with infrastructure design. In the case of residential and commercial buildings, for example, construction codes and standards are needed to maximise the use of natural lighting and to reduce the consumption of energy for air conditioning and heating by including these variables at the design stage.

Codes and standards for residential and commercial buildings can help shape energy use in the future. As an example, the European Union's Building Energy Rating system established 15 categories, ranging from the most efficient, A1, using less than 25 kWh/m^2/year, to the most inefficient, G, using more than 450 kWh/m^2/year. The same rating system is applied to industrial buildings, installations and production equipment. Throughout the LAC region, even though there is a trend towards more efficient, cost-effective and cleaner building construction, the numbers are weak and strong actions are required to promote energy efficiency, and to establish or update energy standards for industry.

Transportation is by far the largest energy consumer in most LAC countries, where the private automobile is often not just a mode of transportation but also a lifestyle choice and status symbol. Thus, the size of the private car fleet is generally increasing, mostly among higher-income groups, while the majority have only access to inadequate public transportation systems based on buses. Only a few cities have good mass transportation systems, although they are usually inadequate to meet the demand. The lack of urban planning and irrational transportation systems are common issues in the region. Thus, major changes in transportation infrastructure and equipment are the most important challenges facing LAC countries if they are to improve the quality of services and ensure more rational energy consumption.

Looking at the international picture, including the LAC countries, over the last decade, there is a clear link between the number of vehicles per thousand inhabitants and GDP per capita. Generally, the size of the vehicle fleet tends to increase as GDP per capita grows, reaching saturation in many high-income countries, averaging around 400–500 vehicles per thousand inhabitants, and even around 800 vehicles per thousand in the United States. The real issue is not the increasing numbers of vehicles as such, but how much each vehicle is used (km per year), and the type and efficiency of public transportation available for people (only buses or in combination with guided traffic systems) and freight (only by trucks or a combination of railways and other modes of transportation).

Thus, a better transportation infrastructure may help to improve energy efficiency and quality of life. Regional infrastructure and energy transportation systems, for example, may help to:

- allocate otherwise 'locked-in' resources;
- reduce transportation costs;
- reduce overall system costs;
- increase regional trade;
- develop long-term commitments; and
- lock-in dependencies.

4.2 *Energy Production and Transmission Infrastructure*

Each country or region builds an energy infrastructure to supply the energy required by the economy. Each infrastructure will be shaped by the energy resources available, by the regional/local population distribution and the location of productive activities, among other factors. There is a relationship between energy supply systems – such as electricity and gas networks for residential, commercial and industrial uses, or district heating systems such as those used in Europe – and end-use infrastructures such as those for transportation, residential and commercial buildings, and urban planning. Taking into account energy considerations at the design stage of urban planning and infrastructure projects can reduce energy demand, costs and the environmental impacts.

4.2.1 Power Generation

The power generation infrastructure that will be needed to meet demand will depend on the energy sources used, whether available locally or imported, first using renewable primary resources such as hydro or geothermal, and then complemented with thermal generation technologies and fuels. Thus priority is given to renewable, clean energy sources, and then non-renewable primary or secondary energy sources, in order of increasing fuel costs and emissions (see Flavin and Aeck 2005; Goldemberg and Lucon 2009).

Looking at the historical trends in the LAC region, this is the situation at the aggregate level. The share of hydroelectric power in the total generation capacity has been stable over the last four decades. This means that hydropower development has been at least proportional to the increase in the total installed power generation capacity. The observed increase in gas-fired power plants (see Fig. 9) has to a large extent been the result of natural gas developments in some gas-producing and importing countries.

The total power generation capacity in the LAC region has increased eightfold, from about 32,000 MW in 1970 to 250,000 MW in 2009. Nuclear power development started in 1974, when the first plant started operation in Atucha, Argentina (340 MW initially). As new plants were built in Argentina, Brazil and Mexico, the regional installed capacity rose to 4,350 MW in 2000. The development of gas-fired power plants has been impressive, following the discovery of natural gas reserves in several LAC countries, and the installed capacity increased by about 600% between 1970 and 2009. However, at the sub-regional level the situation varies. Some of the Southern Cone countries – Argentina, Chile, Paraguay and Uruguay – have been aggressive in the development of efficient combined cycle gas-fired power plants. The installed capacity of gas-fired power plants increased from 400 MW in 1970 to 15,000 MW in 2009, while that of hydropower plants rose from 1,800 MW in 1970 to almost 25,000 MW in 2009.

In Central America, although hydro and geothermal resources are important, in some countries their development has slowed over the last two decades. There was a rapid increase in the installed capacity of diesel oil-fired gas turbines and medium-speed diesel engines using heavy fuel oil, a cheap but polluting fuel. As a result, the installed capacity of gas turbines increased from less than 40 MW in 1970 to 1,200 MW in 2009, while diesel plants, most of them medium speed, increased from 100 MW to 2,500 MW over the same period.

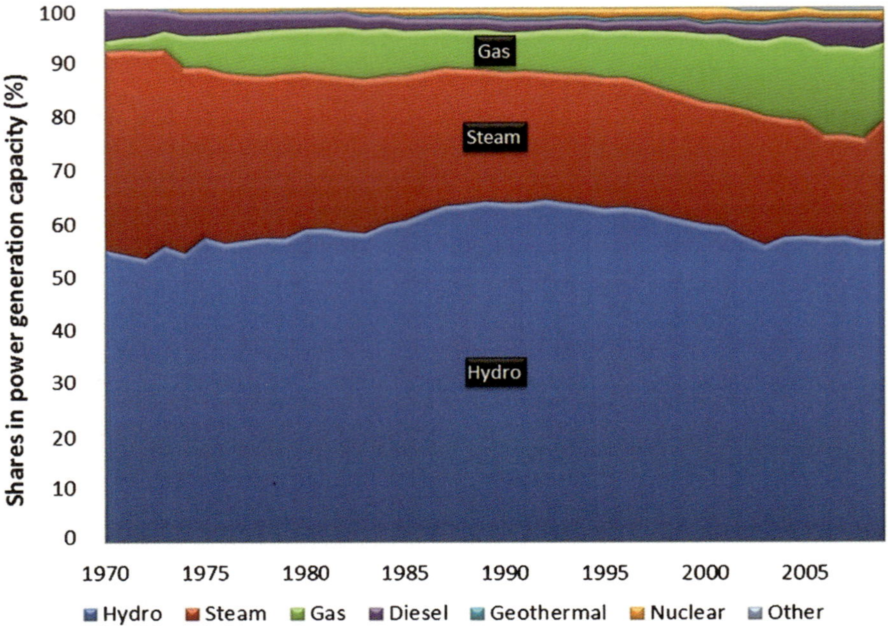

Fig. 9 Power generation structure in the LAC region by type of plant, 1970–2006 (Data source: OLADE 2010)

4.2.2 Power Transmission

The expansion of power transmission networks is directly linked to the population distribution and to the concentration of high-income households in certain economically more developed areas where demand is high. Maps of transmission networks often reproduce the patterns of roads, railways, gas pipelines and other infrastructure systems, all of which converge on the more developed areas within a country or region. But due to widespread poverty, marginalised communities are usually not connected to power grids because of their limited ability to pay. Beyond the populated areas where there are power transmission and distribution networks, there is very poor service coverage, by grid or isolated power plants, especially in rural areas. Even though electricity service coverage has increased significantly – in urban areas it can be above 85–90% – in rural areas there is still room for improvement. In some countries the coverage is low in both urban and rural areas.

Brazil is a good example of the concentration of the power transmission grid in developed areas, especially in the south and southwest, where there is an interconnected system covering the most economically developed areas of São Paulo, Rio de Janeiro and the southern states. The state of São Paulo alone produces nearly half of the total Brazilian GDP. In the northeast, the system covers several states with rapidly growing economies. This is another indication of the strong linkages between the infrastructures for energy, economic development and energy-consuming sectors.

5 Conclusions

This chapter has demonstrated the importance of infrastructure as a determinant variable that shapes the quantitative and qualitative energy demand, and thus the amount and type of investments in energy infrastructure and operations.

There are strong linkages between development, technology and infrastructure, and between energy demand and investments in energy infrastructure expansion. These linkages can be proven on a quantitative basis in the LAC region. There is a high correlation between energy consumption per capita and GDP per capita, which is an indication of the effect of economic growth on energy demand. This linkage materialises through infrastructure and equipment technologies, operational characteristics and actual utilisation. There is also a high correlation between energy consumption and the gross capital stock of the economy, a variable representing the overall investment in fixed capital stock at the macroeconomic level.

The transportation sector is the biggest energy consumer in the LAC region. The major component of energy consumption is directly related to the vehicle fleet, and the number of vehicles per thousand people is strongly correlated with GDP per capita. The number of vehicles and how they are used in turn determine the configuration of the road and highway infrastructure, and the need to introduce more efficient large-scale systems that will shape the future configuration of the public transportation infrastructure. There is a high correlation between industrial energy consumption and industrial GDP, and between household consumption and the macroeconomic aggregate variable of private consumption per capita. Energy intensity indicators suggest that there are inefficiencies at the overall economic level, in transportation and industry, and probably also in households and businesses.

Energy infrastructure is shaped not only by economic development but also by the distribution of the population and concentration of GDP per capita. Power transmission and distribution grids, natural gas pipelines and other forms of infrastructure are oriented to deliver energy products to the major markets. In many LAC countries there are areas with small populations and few economic activities that remain to be developed and require the extension of energy and other infrastructure such as water supply, communications, roads and railways.

Improvements in energy efficiency will lead to lower energy costs and at the same time reduce the investments in energy infrastructure needed to cut greenhouse gas and effluent emissions. Such improvements can be achieved through transportation infrastructure, planned urban development, regulations and standards for buildings and industrial production technologies. To implement these changes will require regulations and standards for the design and construction of such infrastructure, as well as pricing and other regulatory signals to promote the efficient use of both infrastructure and equipment.

Energy planning requires infrastructure planning, and new integrated methodologies. A new methodological approach to energy planning and modelling is needed that is strongly linked to infrastructure planning. This integrated energy–

infrastructure planning should include energy and transportation, water, industry, communications and other sectors, for which information in databases will be crucial. It is essential to open up new lines of investigation focusing on integrated energy–infrastructure modelling, involving energy and the major consuming sectors.

References

Calderón, C., & Servén, L. (2004). *The effects of infrastructure development on growth and income distribution* (Policy research working paper no. 3400). Washington, DC: World Bank.

ECLAC (UN Economic Commission for Latin America and the Caribbean). (2010). Social and economic indicators data base. Santiago: Latin America and Caribbean Statistics (CEPALSTAT)/ ECLAC. http://websie.eclac.cl/sisgen/ConsultaIntegrada.asp

EIA (Energy Information Administration). (2010). *International energy outlook 2010* (Report No. DOE/EIA-0484). Washington, DC: IEA/US Department of Energy. www.eia.gov/oiaf/ieo/index.html

Flavin, C., & Aeck, M. H. (2005). *Energy for development: The potential role of renewable energy in meeting the Millennium Development Goals*. REN21 Renewable Energy Policy Network. Washington, DC: Worldwatch Institute.

Goldemberg, J., & Lucon, O. (2009). *Energy, environment and development* (2nd ed.). London: Earthscan.

Gomelsky, R., Figueroa. F., & Dobrusin, A. (2009). *América Latina y el Caribe, Prospectiva Energética de América Latina y el Caribe, Escenarios Energéticos al 2032* (Vols. I–V). Quito: Latin America Energy Organization (OLADE).

Heaps, C. (2008). *LEAP, Long-range energy alternatives planning system: An introduction to LEAP*. Boston: Stockholm Environment Institute (SEI) US Center.

IAEA (International Atomic Energy Agency). (2006). *Model for analysis of energy demand (MAED-2): User's manual* (Computer manual series no. 18). Vienna: IAEA.

IAEA (International Atomic Energy Agency). (2009). *IAEA tools and methodologies for energy system planning and nuclear energy system assessments*. Vienna: IAEA.

IEA (International Energy Agency). (2009a). *World energy outlook 2009*. Paris: IEA.

IEA (International Energy Agency). (2009b). *Key world energy statistics 2009*. Paris: IEA.

OLADE (Latin American Energy Organization). (2010). *Energy-economy information system*. Quito: OLADE. http://siee.olade.org/siee/default.asp

Energy Efficiency for Development

Chui-Chui Flora Kan

Abstract Improved access to low-cost energy is a key factor in the battle to alleviate poverty and achieve basic development goals. The outcomes of combined development and energy efficiency programmes in Senegal, Nicaragua, China and Ukraine demonstrate that better energy efficiency can also contribute to health and education, particularly among women and children in developing countries. Improving energy efficiency can significantly increase the returns on rural economic investments and can contribute to energy security and support sustainable economic growth.

1 The Importance of Energy Efficiency to Development

An important first step in addressing this topic is to find a sound definition of energy efficiency. The World Energy Council offers the following: *'Energy efficiency improvements refer to a reduction in the energy used for a given service (heating, lighting, etc.) or level of activity'* (WEC 2004, p. 3). This definition specifically refers to the provision of services, and not to restrictions or reductions in their use. This is the definition I have chosen because, with respect to socio-economic development, pursuing energy efficiency should not be viewed as a way to reduce the use of energy-intensive services. On the contrary, improving energy efficiency facilitates the use of services that rely on electricity by reducing running costs.

Improving energy efficiency is important for development. As observed in previous chapters (Brew-Hammond 2012; Pachauri 2012), 1.6 billion people worldwide do not have access to electricity, and 2.5 billion people rely on the use of

C.-C.F. Kan (✉)
UNDP China End Use Energy Programme, Programme Management Office,
Guohong Building B, Xicheng District Beijing, 100038, China

F.L. Toth (ed.), *Energy for Development: Resources, Technologies, Environment*, Environment & Policy 54, DOI 10.1007/978-94-007-4162-1_9,

wood, dung or agricultural residues for heating and cooking (IEA 2006a). My contention is that enhanced access to low-cost energy is a key factor in the battle to alleviate poverty and achieve basic development goals.

How can access to energy help development and alleviate poverty? My response, perhaps surprisingly for a physicist, comes from the Bible: 'Let there be light' (Genesis 1:14). Having artificial light means that people can read and study in the evenings, and initiatives such as 'One Laptop per Child' project can be more effective in helping children in developing countries to acquire knowledge. Access to electricity means that entrepreneurs can use electrical tools and other equipment to produce quality goods in quantity, thus increasing their incomes far more rapidly than they could ever do with manual tools. Electricity powers refrigerators that extend the shelf life of food, as well as radios, televisions, computers and mobile phones, making communication services possible. With access to the internet, people can learn how much things cost elsewhere and use that knowledge to buy and sell goods at realistic and competitive prices. Such knowledge can also help traders avoid the risks posed by factors beyond their control, such as a collapse in the market elsewhere in the country or the world.

Meanwhile, having access to electricity can save time. The majority of the 2.5 billion people who have no access to electricity must collect firewood and agricultural residues to meet their basic energy needs, and that takes time. That burden falls mainly on women and children.

This chapter presents four examples of projects in developing and emerging economies that are addressing some of the issues associated with energy efficiency and energy access.

2 Four Energy Efficiency Projects

2.1 *Example 1: Senegalese Rural Electrification Agency*

Most energy experts now accept that access to energy can serve to alleviate poverty. But is it incongruous to try to improve both energy access and demand-side management? The Senegalese Rural Electrification Agency (ASER) has demonstrated that it is possible to do both. ASER has a mandate to provide access to electricity to 50% of the population, or about 3.8 million people, and has made steady progress towards reaching that goal. By 2000 the agency was providing electricity to 8% of the population, and by 2006 that figure had risen to 16%. ASER expects to reach the target of 50% by 2012 (UNDP 2008). As a result there will be many more people with access to electricity, as well as to new energy-efficient appliances.

A special aspect of ASER is that alongside providing electricity access to new consumers, it is also implementing the Rural Energy-Efficient Lighting Programme, which aims to help households minimise electricity use from the start, through the use of energy-efficient light bulbs and motors. This initiative supports the distribution of some 1.5 million compact fluorescent lamps (CFLs) to newly electrified houses and other buildings, funded by the Community Development Carbon Fund of the Clean Development Mechanism (CDM), with the support of the World Bank (CDCF 2009; World Bank 2008a, b). The CFLs are being distributed in order to forestall any demand for inefficient, short-lived incandescent light bulbs. ASER also supports a recycling programme so that used CFLs are collected and disposed of safely to prevent any possible mercury pollution.

The increasing availability of electricity will naturally result in the use of more electrical appliances. In particular, the programme anticipates that the many owners of internal combustion motors will want to replace them, so it also provides subsidies for purchases of energy-efficient electric motors. It is estimated that between 2008 and 2013, the use of CFLs rather than fluorescent light bulbs will save about 45 gigawatt-hours (GWh) of electricity per year. For both the new consumers in the rural areas of Senegal and electricity providers, this is a significant achievement.

2.2 *Example 2: The Pro-Tortilla Programme, Nicaragua*

Energy efficiency can improve health, raise living standards and allow additional time for education. The Pro-Tortilla Programme (which won an Ashden Award for Sustainable Energy in 2003) operates in the departments of Managua, León and Granada in western Nicaragua. Every month, residents of these three departments consume about 30 million tortillas, most of which are produced by 4,000 household businesses known locally as 'Mom and Pop' stalls. These enterprises are very important for the local economy and represent a major source of income for a significant number of people, especially considering that the average unemployment rate is 50%.

The stoves traditionally used to produce tortillas are open-fire stoves without chimneys, which produce a lot of smoke. Indoor air pollution is substantial, leading to adverse health effects such as respiratory problems. As an alternative, the Pro-Tortilla Programme has developed the 'ecostove', which uses about 50% less wood, produces much less smoke and generates about 35% less CO_2 than traditional stoves. The programme has had substantial impacts – producers can prepare many more tortillas at a lower cost without damaging their health. Perhaps most important, they need to spend far less time collecting firewood, allowing them to devote more time to their children.

2.3 *Example 3: Rural Tourism Development Project, Pinggu, China*

The Pinggu Solar-Heated Housing for Rural Tourism Development project, implemented by Beijing municipality, has demonstrated that energy efficiency can complement renewable energy investments. Located about 70 km northeast of

Beijing, with its population of 17 million, Pinggu district has an enormous potential as a tourist destination.

In its 11th five-year plan, the Chinese government set out an economic development strategy for the rural areas for the period 2006–2010. Its targets included raising average rural per capita net incomes from 3,255 renminbi (RMB; US$369) per year to 4,110 RMB (US$644) and increasing the proportion of incomes from non-agricultural sources (People's Daily 2006). At the same time, in its own 11th five-year plan, Beijing municipality launched the New Socialist Countryside Initiative, a programme that aims to integrate agriculture and rural initiatives with China's modernisation drive. It would also promote the application of sustainable energy technologies in the rural areas. The municipality's economic development targets include increasing average rural per capita net incomes from 7,860 RMB (US$1,231) per year in 2006 to 10,000 RMB (US$1,566) in 2010, and achieving *XiaoKang* (the Chinese equivalent of the Millennium Development Goals), i.e. doubling net per capita incomes for 90% of rural households from 2000 levels.

Beijing municipality's strategy for rural development is to support the establishment of tourism, and to encourage city dwellers to take short breaks and spend money in the rural areas. The municipality selected 13 less developed villages in 10 districts for the piloting of the New Socialist Countryside Initiative (Invest Beijing 2007). Between 2006 and 2008, the municipality constructed 228,000 m^2 of new buildings complete with solar heating systems, 90% of which were located in Pinggu district.

Pinggu district is an attractive area, with mountains and lakes and good transport and communication links. Surrounded by hills on three sides, the district enjoys plentiful sunshine throughout the year. Pinggu is a major supplier of agricultural produce to Beijing, and is renowned for its peaches. Pinggu is one of 11 new towns on the outskirts of Beijing.

In 2004, Pinggu district investigated eight heating options: district heating, electric heating, solar, hanging oven bed, bio-digesters, biomass stoves, natural gas and wind power. It then assessed the circumstances of Pinggu, namely its transport links, abundant sunshine and scenic surroundings, recognising the need to preserve air quality and other resources. The district council designated Gua Jia Yue village as a demonstration project, which would include installing solar heating alongside additional building conservation measures.

Existing methods of domestic heating in Pinggu were rather primitive, including heated oven beds and other forms of heating with very low thermal efficiency and supplying low-quality heat. Standards of hygiene were also very poor. The only fuels used to heat water were coal or firewood, which meant that hot water supplies were sporadic. These living conditions were far below the standards expected by city dwellers, and presented a fundamental barrier to the development of rural tourism.

Houses in rural areas such as Pinggu are generally poorly insulated, with very low indoor temperatures and limited hot water supplies in winter, while temperatures are very high both indoors and outdoors in summer. Over 90% of rural households maintain an indoor temperature of less than 15°C in winter, and only 2% keep

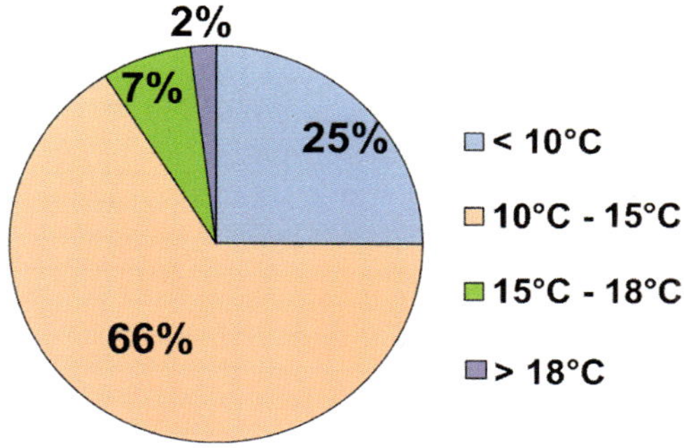

Fig. 1 Average indoor temperature of rural homes in Beijing in winter (Source: Zheng et al. 2008)

the temperature above 18°C (see Fig. 1). For city dwellers, that would not make for a very enjoyable holiday destination!

Today, there are 1,508 solar-heated houses in Pinggu where the temperature remains above 18°C in winter and below 26°C in summer. To put this in context, the government of China requires that all public buildings maintain a maximum indoor temperature of 26°C in summer, using air conditioning. This number is significant as this is the comfort level that Beijing residents are used to. For the Gua Jia Yue demonstration project, insulation was included in the design from the start, alongside solar heating and hot water systems. The municipality financed the construction, with grants of 40%, and interest-free loans of up to 43% of the total investment, repayable over 8 years. This was a significant innovation because peasants traditionally find it very difficult to obtain credit. The 40% grant covers not only the installation of energy-efficient equipment, but also the additional infrastructure and facilities required to facilitate tourism, including extra rooms that households would not normally require.

The UNDP China End Use Energy Efficiency Project (UNDP 2003, 2007) monitored the impacts of the Gua Jia Yue pilot project, including hot water and heating supplies, the energy savings and CO_2 emissions over the course of 1 year. We found that the present value of the subsidies, after deducting additional tax revenue that the local government will receive from tourism, is more than US$27 million. However, the net income (i.e. excluding income tax and operating costs) from this investment in rural tourism will be US$124 million over 15 years. We looked at 15 years because that is the approximate life span of a solar heating system. The energy efficiency improvements would last about 25 years and when this saving is included in the calculation, the present value of the net income is US$163 million over those 25 years. I think that this higher figure should be considered because once the project is established, households should be able to afford to replace the solar heating systems themselves after 15 years. Looking at this investment in terms of CO_2 emissions over 25 years, approximately 740,000 tonnes of CO_2 emissions will be avoided, at a cost of US$37 per tonne (US$27 million in total).

For the residents of Gua Jia Yue, the pilot project has so far generated additional incomes averaging 6,000 RMB (US$930) per month. Living and hygiene standards have also improved significantly. Indeed, there is anecdotal evidence that girls across the region are keen to marry bachelors from Gua Jia Yue village, no doubt attracted by the prospect of a warm home with hot showers and a steady income.

Commercial banks are reluctant to lend to projects with low returns or to borrowers with few opportunities to generate incomes to repay their loans. Also, energy conservation measures alone are sometimes difficult to justify. Traditional houses have little or no insulation, so rural inhabitants tend not to heat them. If no heating is used, it is not possible to say that installing energy conservation measures actually saves or conserves energy. The Pinggu project has demonstrated that energy efficiency really can improve the economic returns on investments in rural development.

2.4 Example 4: Ukraine Energy Efficiency Programme

Ukraine is a major net importer of oil and gas, and its economy remains one of the most energy-intensive and inefficient in Europe. The Ukraine Energy Efficiency Programme (UKEEP), implemented by the European Bank for Reconstruction and Development (EBRD), aims to demonstrate that improving energy efficiency can enhance energy security. In Ukraine, the energy intensity of GDP is very high, at about 0.5 tonne of oil equivalent (toe)/US$1,000 (at 2,000 prices, converted at purchasing power parity; IEA 2006b, p. 80). For comparison, the energy intensity of GDP in the European Union is 0.18 toe/US$1,000, and the world average is 0.21 toe/US$1,000. In the 4 years from 2005 to 2008, Ukrainian electricity prices rose by about 110%, or 20% per year, and imported gas prices increased by 280%, or 40% per year.

According to the EBRD (2007, p. 64), about 45% of the gas consumed in Ukraine, is used in district heating systems (see Fig. 2 for data for earlier years). Some 30% of the heat is lost due to defective pipework, and 50% of the heat supplied to buildings is wasted due to poor insulation, outdated equipment and inefficient practices. If all the energy conservation measures applicable to district heating systems and to the building sector were implemented, the potential savings could amount to 223 Mtoe, and overall gas consumption could be reduced by 25–30%. Table 1 illustrates the potential to reduce gas consumption through energy efficiency improvements in district heating systems and buildings.

This example shows that energy efficiency improvements can enhance energy security by reducing the demand for gas and thus the reliance on energy imports.

3 Conclusions

Four important lessons can be drawn from these examples. First, access to energy is essential for development, and new users can use energy more efficiently than existing ones. Second, energy efficiency can contribute to health and education, particularly

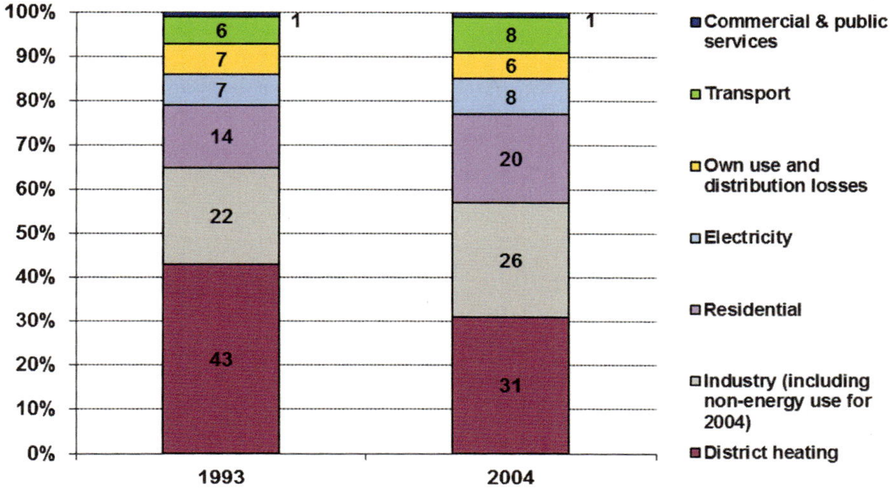

Fig. 2 Natural gas consumption in Ukraine, by sector, 1993 and 2004 (Source: IEA 2006b)

Table 1 Gas consumption in Ukraine before and after energy efficiency measures

Gas used in	Percentage of total gas consumption (%)	Heat losses due to inefficiency (%)	Efficiency gains with measures	Percentage of total gas consumption with efficiency measures (%)
District heating	45	30	13.5% through improved pipework	31.5
Buildings (based on efficient heat supply)	31.5	50	15.75% through improved building insulation	15.75
Others	55			

among women and children, in developing countries. Third, energy efficiency improvements can significantly increase the returns on rural economic investments. Fourth, energy efficiency can contribute to energy security and support sustainable economic growth.

The demand for energy services will undoubtedly grow, especially in the developing world. Its impact will be to reduce poverty and drive development, resulting in a corresponding rise in energy use. I believe we can taper this growth using energy efficiency measures and, if we do that, there will be many winners, in terms of better health, higher living standards, lower pollution, lower energy costs and more rapid development.

References

Ashden Awards for Sustainable Energy. (2003). *The Pro-Tortilla programme: Modernisation of household tortilla businesses in Nicaragua with the 'Ecostove'*. Case study. www.ashdenawards.org/winners/prolena. Accessed 18 May 2010.

Beijing, I. (2007). *New socialist countryside initiative* (Vol. 8). Beijing: Beijing Municipal Economic Information Centre.

Brew-Hammond, A. (2012). Energy: The missing Millennium Development Goal. In F. L. Toth (Ed.), *Energy for development: Resources, technologies, environment* (pp. 35–43). Dordrecht: Springer.

CDCF (Community Development Carbon Fund). (2009). *Assessment of community benefits and sustainable development*. Washington, DC: World Bank.

EBRD (European Bank for Reconstruction and Development). (2007). *Strategy for Ukraine*. London: EBRD.

IEA (International Energy Agency). (2006a). *World energy outlook 2006*. Paris: OECD/IEA.

IEA (International Energy Agency). (2006b). *Ukraine: Energy policy review 2006*. Paris: OECD/IEA.

Pachauri, S. (2012). Demography, urbanization and energy demand. In F. L. Toth (Ed.), *Energy for development: Resources, technologies, environment* (pp. 81–94). Dordrecht: Springer.

People's Daily. (2006, March 8). *China sets new targets in five-year plan*. http://english.peopledaily.com.cn/200603/07/eng20060307_248647.html

UNDP (United Nations Development Programme). (2003). *Project document: China End Use Energy Efficiency Programme*. Beijing: UNDP.

UNDP (United Nations Development Programme). (2007). *Project summary: China End Use Energy Efficiency Programme (EUEEP)*. Beijing: UNDP. 2007–09.

UNDP (United Nations Development Programme). (2008). *The Bali road map: key issues under negotiation*. New York: UNDP.

WEC (World Energy Council). (2004). *Energy efficiency, a worldwide review: Indicators, policies, evaluation*. London: WEC and French Environment and Energy Management Agency (ADEME).

World Bank. (2008a, September 4). *A first for Senegal as CDM agreement signed for efficient lighting program in rural areas*. Press release. http://go.worldbank.org/N598PN6Y60

World Bank. (2008b). *Carbon finance for sustainable development 2008*. Washington, DC: World Bank.

Zheng, Z. L., Li, Y. H., & Yang, X. D. (2008). Rural building energy conservation research of Beijing municipality. *Building Science Journal, 24*(4), 9–14.

Externalities in the Global Energy System

Stefan Hirschberg

Abstract This chapter addresses externalities in the global energy system and introduces a methodology for assessing environmental impacts and external costs. The health effects due to air pollution and global warming dominate current estimates of external costs. These costs may be substantial and should affect investment decisions, but there are large variations between technologies and locations. Consistent comparisons of technologies must clearly differentiate between past and future technologies. Good technologies, including advanced fossil fuels, have rather low external costs due to air pollution, while climate change contributions may be very high but are subject to very large uncertainties. The external costs of nuclear and renewables are low. The social aspects of energy systems are represented only to a limited extent in current estimates of external and total costs. The inclusion of broader social factors in the multi-criteria decision analysis framework favours renewables and is a challenge for future nuclear.

1 Introduction

This chapter discusses externalities in the global energy system. Section 2 begins with the basics: what are externalities? Section 3 introduces a methodology for assessing environmental impacts and external costs. Section 4 presents examples

---

S. Hirschberg (✉)
Laboratory for Energy Systems Analysis, Paul Scherrer Institut,
5232 Villigen PSI, Switzerland
e-mail: stefan.hirschberg@psi.ch

F.L. Toth (ed.), *Energy for Development: Resources, Technologies, Environment*, Environment & Policy 54, DOI 10.1007/978-94-007-4162-1_10,

of assessments of external costs, and discusses total costs, which are the sum of internal and external costs, with examples of their uses in the context of cost–benefit analysis. Estimates of external costs, like any other complex method, have major limitations. Section 6 addresses some of these limitations and introduces a complementary, aggregating assessment method for energy systems. This method, multi-criteria decision analysis, is broader in scope than the external costs framework. Finally, Sect. 7 presents some conclusions about externalities.

2 Externalities

In this chapter, externalities are defined as changes in welfare generated by a given activity that are not reflected in market prices. There are positive externalities that yield benefits, as well as negative externalities that result in costs. A cost is external when it is not paid for by those who have generated it. This means simply that one group may cause problems for other groups, but is not held responsible. The focus here is on costs rather than benefits because, from the point of view of consumers, 1 kW-hour (kWh) of electricity delivered at the time it is needed has basically the same benefit, the same value, regardless of how it is produced. So with regard to specific benefits, the difference between the various generating technologies is very small and normally concerns rather special aspects (for example recreational uses of hydro dams).

Someone has to pay for negative externalities, and this is usually done by all of us, by the tax payers. Thus, for increasing the efficiency of the economy, such costs should be reduced and paid for by those who generate them. In this context, the so-called 'polluter pays' principle (OECD 2001, 2005; Tol 2006) is the appropriate way to internalise external costs, but today we are far from internalising all the external effects to which the public and ecosystems are exposed.

Air pollution is often the dominant factor in environmental external costs (Jamal et al. 2004; Karimzadegan et al. 2008; WHO 2010). It results, for example, in increased hospital admissions due to respiratory illnesses, leading to very measurable, concrete costs, including the increased costs of healthcare. At the same time, people who are affected by air pollution are not productive while they have health problems. There are also other costs that are much more difficult to monetise objectively, such as the pain and suffering experienced by patients under such circumstances, and the emotional effects on friends and relatives.

Another example is water pollution (UN WWAP 2003, 2006; UNEP 2010), which leads, for example, to the loss of fish stocks, reduced recreational opportunities, commercial losses and impacts on biodiversity that are often difficult to quantify. A third example that everyone has experienced is traffic congestion, which results in lost time, lost productivity and also increased stress (Mayer and Sinai 2003; Parry et al. 2007; Richardson and Bae 2008; Currie and Walker 2009; Sen et al. 2010).

3 Methods for Assessing Environmental Impacts and External Costs

Table 1 is taken from a recently completed project, New Energy Externalities Development for Sustainability (NEEDS), a European study of externalities involving the Paul Scherrer Institute (PSI) and 65 other participating organisations. The table shows a number of environmental externalities, including impacts on human health, that can now be quantified, and which dominate current estimates of external costs. The table includes human health effects caused by a number of pollutants, such as mortality (reduced life expectancy) and morbidity, i.e. illnesses that can be relatively trivial but unpleasant, such as a cough, or can be very severe and expensive for society, such as chronic bronchitis.

In addition to the human health impacts, there are also effects on buildings, crop yields and the difficult issue of global warming. There are also effects that are difficult to monetise, including the loss of amenity and effects on ecosystems and land use.

Figure 1 shows how these costs are estimated. A power plant in a specific location is identified; this power plant produces emissions, both directly as well as indirectly, through its energy and material inputs. Here only the power plant is shown, but the whole energy chain is examined, from the extraction, processing and transportation of fuels, to electricity generation and, finally, waste management. All phases in the evolution of this facility are considered, including construction, operation and dismantling. We then estimate the total emissions of major pollutants: SO_x, NO_x, particulates and ammonia. Ammonia is very important because SO_x and NO_x, when they react chemically with ammonia, produce secondary particulates that actually cause most of the health problems due to air pollution.

As the next step, the dispersion of pollutants is modelled (Holmes and Morawska 2006; Stein et al. 2007; Marshall et al. 2008; Steyn and Rao 2010). This is based on large simulation models whose outputs are estimates of the increases in pollutant concentrations. According to epidemiological studies there is a clear correlation between increased pollutant concentrations and the effects on health. Even if a single plant is selected for this analysis, an area of several million square kilometres needs to be covered because the particulates can travel very large distances, even from one continent to another. The next step is to estimate the physical impacts in terms of human health effects and crop losses. In the context of mortality, we calculate the years of life lost (YOLL) due to air pollution per kilowatt-hour of power produced (YOLLs/kWh). The final step, which is very useful for cost–benefit analysis, is the monetary valuation of external costs in euros or dollars per kWh.

There are several tools for assessing externalities. The main tool used in Europe is the EcoSense model, which was developed for Europe, Russia, Brazil and China. These are all highly specific versions with enormous databases, with data covering not only emissions but also specific characteristics of each cell in the mesh shown in Fig. 1. These include detailed information on the receptors exposed to air pollution, i.e. human beings, flora and fauna, as well as meteorological conditions such as winds and precipitation.

Table 1 Externalities currently included in estimates of environmental costs (NEEDS) (Source: Friedrich 2007)

Impact category	Pollutant/burden	Effects
Human health: mortality	PM_{10}	Reduced life expectancy due to short- and long-term exposure
	SO_2, O_3	Reduced life expectancy due to short-term exposure
	Benzene, BaP, 1,3-butadiene	Reduced life expectancy due to long-term exposure
	Diesel particulates, radioactivity, HM, noise	Reduced life expectancy due to short-term exposure
	Accident risk	Increased risk of fatalities due to traffic and workplace accidents
Human health: morbidity	PM_{10}, O_3, SO_2	Respiratory hospital admissions
	PM_{10}, O_3	Restricted activity days
	Benzene, BaP, 1,3-butadiene	Congestive heart failure
	Diesel particulates, radioactivity	Cancer risk (non-fatal)
	PM_{10}	Increased hospital admissions of patients with cerebrovascular and respiratory problems, chronic bronchitis, chronic coughing in children, asthmatics
	Noise	Myocardial infarction, angina pectoris, hypertension, sleep disturbance
	Mercury	Lower IQ of children
	O_3	Asthma attacks, symptom days
	Accident risk	Risk of injuries due to traffic and workplace accidents
	Other heavy metals	Diverse health impacts
Building materials	SO_2, acid deposition	Ageing of galvanised steel, limestone, mortar, sandstone, paint, rendering, and zinc for buildings
	Combustion particles	Soiling of buildings
Crops	SO_2	Changes in yields of wheat, barley, rye, oats, potato, sugar beet
	O_3	Changes in yields of wheat, barley, rye, oats, potato, rice, tobacco, sunflower seed
	Acid deposition	Increased need for liming
	N, S	Fertilising effects
Global warming	CO_2, CH_4, N_2O	Worldwide effects on human mortality and morbidity
		Impacts on coastal areas, agriculture, energy demand and economies due to temperature change and sea level rise
Amenity, ecosystems and land use	Noise	Amenity losses due to noise exposure
	SO_2, NO_x, NH_3	Eutrophication, acidification of ecosystems 'PDF' of species

Notes: *HM* heavy metal, *PM* particulate matter, *PDF* probability density function

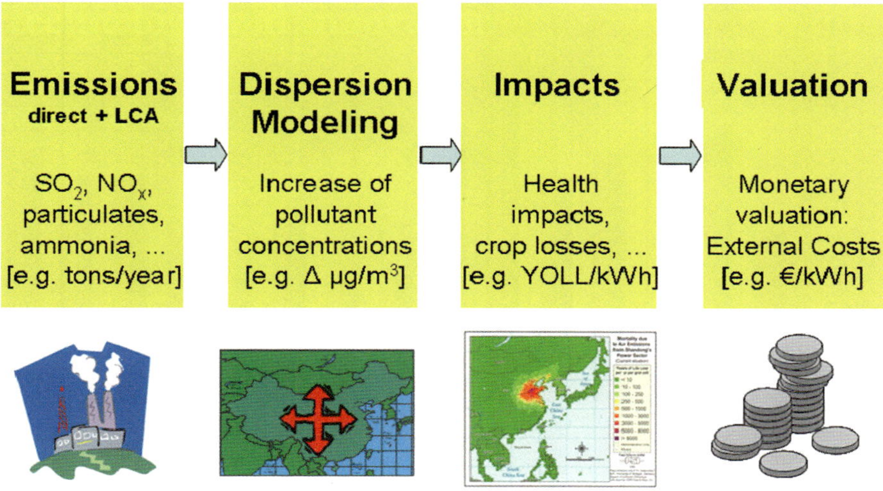

Fig. 1 Environmental impact assessment – the impact pathway approach (Note: *YOLL* years of life lost)

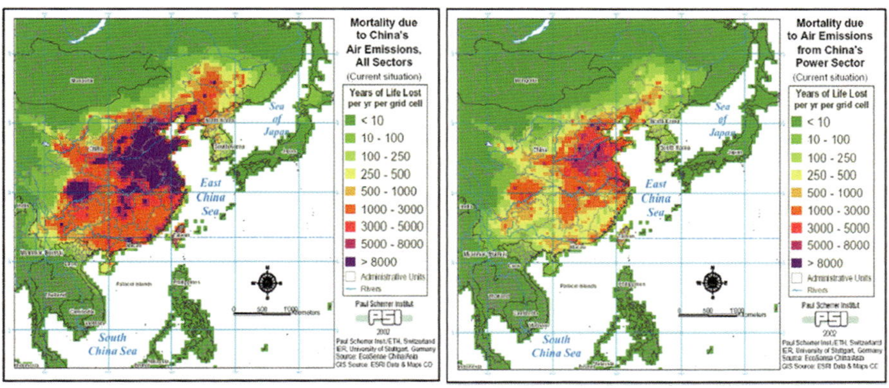

Fig. 2 Mortality in China due to air pollution from all sectors (*left*), and from the power sector (*right*) (Source: Hirschberg et al. 2003)

4 Examples of Impact and External Cost Results

Figure 2 shows the results of the simulations for China, in terms of the overall mortality due to air pollution (Hirschberg et al. 2003). For each grid cell (roughly 50 × 50 km), the years of life lost per year and per grid cell are shown; the darkest cells represent the worst loss of life. The map on the left shows the emissions from all sectors, and the one on the right emissions from the power sector. The troubling aspect is that, at the time of the analysis, the power sector contributed about 25% of

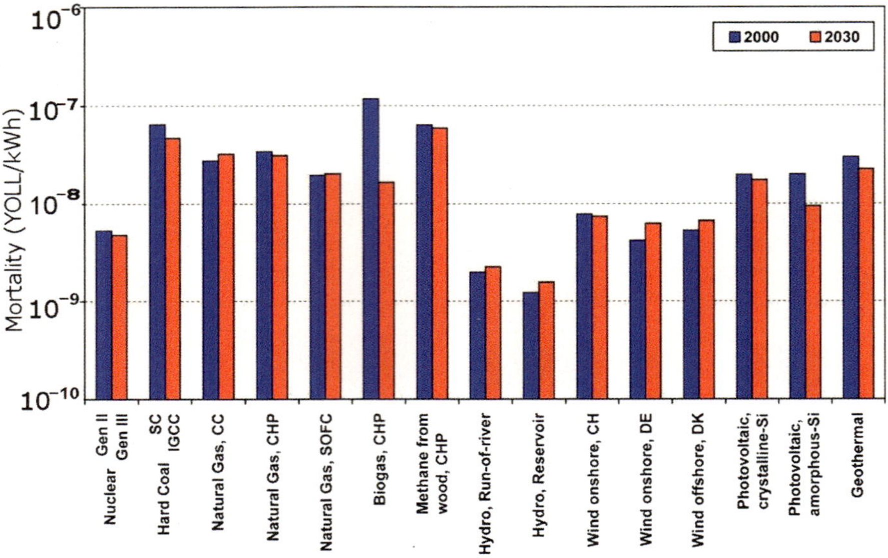

Fig. 3 Mortality due to the normal operation of various energy technologies in Switzerland, 2000–2004 and 2030 (Source: Hirschberg et al. 2008a; Note: *YOLL* years of life lost)

the overall health and environmental damages due to air pollution, it is the one sector that has been growing in China at a very high rate since then. The decisive question is whether China's emissions will continue to increase at the same rate as in the past, or whether major steps will be taken to limit them.

The darkest cells in Fig. 2 represent more than 8,000 years of life lost. Specifically, if we examine 1 GW power plants without filters (many plants in China do not have SO_x filters) located in or close to densely populated areas, each plant is responsible for somewhere between 20,000 and 30,000 years of life lost. For the whole of China, this amounts to roughly ten million years of life lost per year, or about one million premature deaths each year due to air pollution. These are socially highly significant numbers.

Figure 3 shows a very different case, mortality due to the normal operation of very good existing and advanced future technologies, mainly in Switzerland (Hirschberg et al. 2008a). The numbers are much lower than for China, not only because of lower emissions but also because of the lower population density. All of these technologies are of interest to Swiss utilities, and have been evaluated by PSI as a part of an analysis of electricity supply portfolios for the current situation (2000–2004) and in the future (2030). Surprisingly, the largest health effects come from current biogas combined heat and power (CHP) plants, and are actually higher than those due to coal or natural gas, which are represented by excellent, clean technologies. The results for CHP plants also depend significantly on the choice of method of allocating impacts to electricity. Here the 'exergy' allocation method has been applied. (The exergy of an energy or material quantity measures its usefulness

Table 2 Examples of valuation for monetizing external costs

Health impact	EU current[a] (€2,000)	EU 1999[b] (US$2,000)	China[c] (US$2,000)
Value of a statistical life	3,000,000 (infant mortality risk)	3,000,000	440,000
Acute years of life lost/reduced life expectancy (per YOLL)	60,000	110,000	15,710
Chronic years of life lost/reduced life expectancy (per YOLL)	40,000	110,000	15,710
Restricted activity days (per case)	130	116	17
Chronic bronchitis (per case)	200,000	178,000	25,400

Sources: [a]Desaigues et al. (2006), [b]European Commission (1999), [c]Hirschberg et al. (2003)

or quality. Exergy-based economic methods combine exergy and economics to achieve thermodynamic and economic objectives by assigning costs and/or prices to exergy-related variables.) Hydropower shows the best performance, followed by wind and nuclear. The health effects due to wind and photovoltaics are not due to direct pollutant emissions from power plants. They are based on the life cycle assessment approach that accounts for the material flows and the energy inputs needed for the construction of these facilities and transportation.

Table 2 shows the values used for the final step of monetisation, some of which are controversial. They normally reflect willingness to pay, based on broad surveys in which members of the public were asked how much they would be willing to pay in order to reduce the risks. The numbers may change over time. In the EU, the value of a statistical life is currently €3 million, while the cost of a year of life lost due to chronic illness is €40,000, and of a case of bronchitis €200,000.

Figure 4 shows the external costs of the energy technologies considered in the NEEDS project, most of which again are for excellent technologies used in Western Europe (DLR et al. 2009). The table includes contributions from different components, i.e. regional health impacts, climate change and other non-climate impacts. Estimates of the external costs associated with climate impacts are subject to very large uncertainty, as illustrated by the range in the figure. The external costs of fossil technologies are substantial and, for the higher levels of climate change, actually exceed the costs of generation, i.e. internal or private costs. Nuclear and wind have the lowest external costs. The hydrogen case is based on the production of hydrogen by electrolysis using the current European average electricity mix. The results would be much more favourable if either renewables or nuclear were used to produce hydrogen.

Figure 5 shows the projected external costs of various energy technologies in 2050 (DLR et al. 2009). The future external costs are lower than they are at present partly due to advances in technology, partly due to discounting the cost of damage due to climate change and partly due to structural changes. Hydrogen is produced using renewable electricity. But although the external costs of these advanced technologies are lower, they remain significant, particularly for coal and biomass.

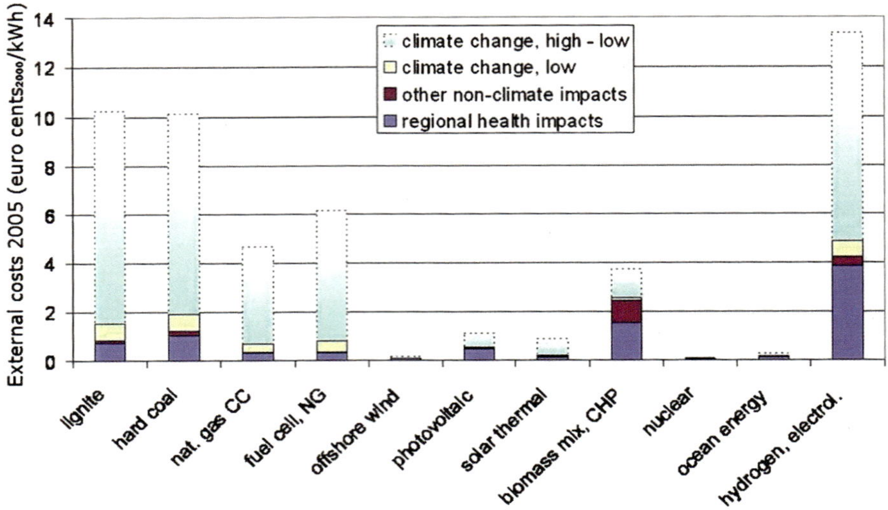

Fig. 4 External costs of various energy technologies in Western Europe in 2005 (Source: Adapted from DLR et al. 2009; Note: *CC* combined cycle, *NG* natural gas, *CHP* combined heat and power, *electrol* electrolysis)

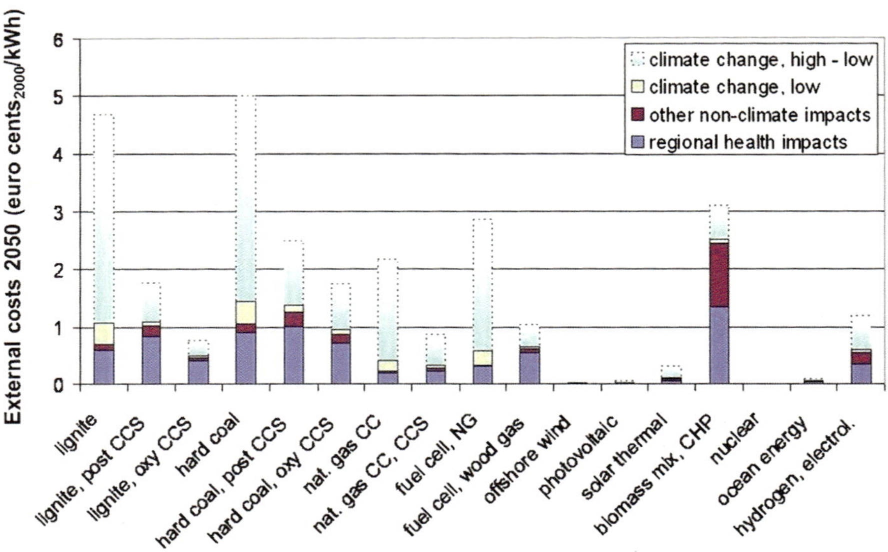

Fig. 5 External costs of energy technologies in Western Europe in 2050 (Source: Adapted from DLR et al. 2009; Notes: *CCS* carbon dioxide capture and storage, *CC* combined cycle, *NG* natural gas, *CHP* combined heat and power, *electrol* electrolysis)

5 Total Costs and Their Limitations

Figure 6 shows the total costs – the sum of external and internal costs – of the electricity generation technologies of interest to Swiss utilities (Hirschberg et al. 2008a; Roth et al. 2009). From one perspective, the total costs are considered to be a measure of the economic and environmental efficiency of energy systems. The results are mostly for Swiss conditions. For nuclear, generation II and III plants were analysed; for GEN III plants the European Pressurized Reactor (EPR) was used as a reference. Nuclear shows the lowest total costs both today and in the future. For solar photovoltaics, cost reductions are achieved due to technological improvements. But under Swiss conditions, solar is still the most expensive technology in 2030 in terms of total cost, due to the large internal costs. Geothermal is very promising. It is worth noting the relative competitiveness of onshore wind in Germany and offshore wind in Denmark, and the large reductions in the total costs of solid oxide fuel cells and of biogas due to the use of improved, more environmentally friendly technologies. However, in the case of Switzerland, the potential of biogas is relatively limited.

In China, the total costs of fossil technologies are on a totally different level. Figure 7 shows the costs of plants in Shandong province, and the external costs far exceed the internal costs (Hirschberg et al. 2003). At the time of the study, all the power generated in Shandong province was based on coal-fired plants without SO_x

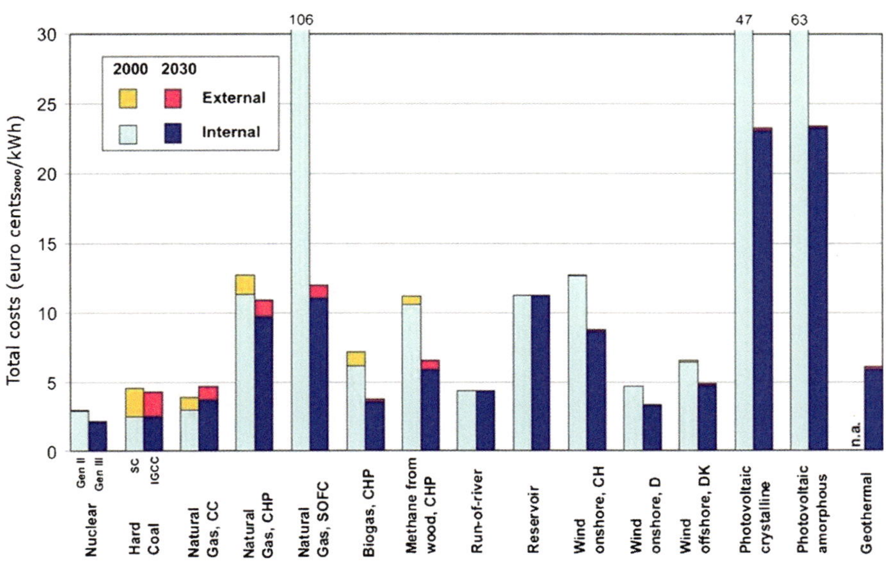

Fig. 6 Total costs of electricity generation options in 2005 and projections for 2030 (Sources: Hirschberg et al. 2008b; Roth et al. 2009; Note: *CC* combined cycle, *CHP* combined heat and power, *SOFC* solid oxide fuel cell, *CH* Switzerland, *D* Germany, *DK* Denmark, *n.a* not applicable)

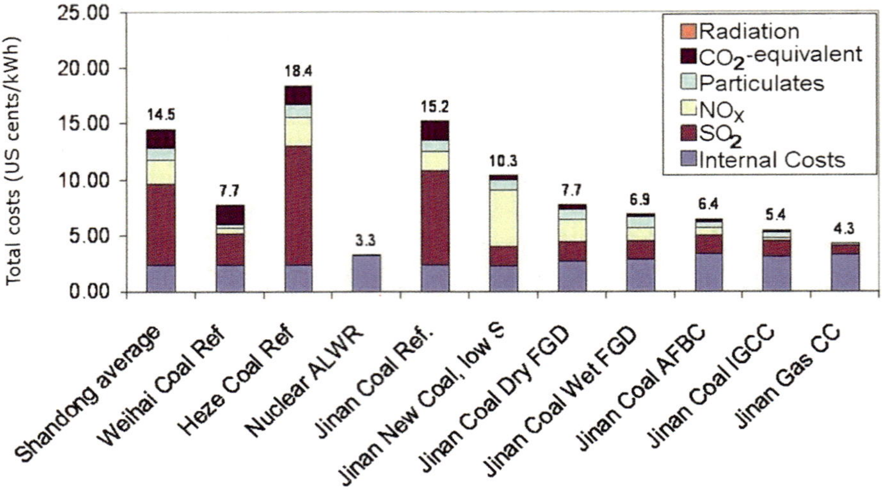

Fig. 7 Total costs of electricity generation in the Chinese province Shandong for technologies used around year 2000 (reference and average) and for technologies that could be implemented before year 2020 (Source: Hirschberg et al. 2003; Note: *Ref* reference, *ALWR* advanced light water reactor, *S* sulphur, *FGD* flue-gas desulfurization, *AFBC* atmospheric fluidised bed combustion, *IGCC* integrated gasification combined cycle, *CC* combined cycle)

and NO_x filters. The figure illustrates the variations in external costs of plants with location. For coal-fired plants on the coast (Weihai), part of air pollution is dispersed in the direction of the sea, and so does not contribute to human health impacts. For those located inland (Heze, Jinan), the effects are much greater. The figure illustrates the technological possibilities for reducing external and total costs: using low-sulphur coal, using filters on power plants, or diversifying by including nuclear or natural gas combined cycle plants. Changing technology has enormous impacts.

Figure 8 shows the results of an analysis of 18,000 electricity supply scenarios for Shandong province, conducted by PSI together with the Massachusetts Institute of Technology (MIT). The results are based on groupings of these scenarios (Hirschberg et al. 2003). For example, one can see the effects in 2020 of expanding the electricity sector by installing filters at all coal plants. The extra cost per year of installing the filters is US$0.5 billion in 2020, which pales in significance compared with the benefit of US$10.2 billion in terms of external costs avoided. Also, if the value assigned to human life is reduced (this is always a controversial issue), even by a factor of 10, there is still a clear benefit in installing more environmentally friendly systems. This effect is even greater when both clean coal technologies and diversification are considered. In the case of Shandong, diversification would mean including natural gas, nuclear and some wind energy, since the province has no hydro resources.

Figure 9 shows the magnitude of worldwide damage due to CO_2 emissions and air pollution, and the effect of implementing environmentally friendly systems

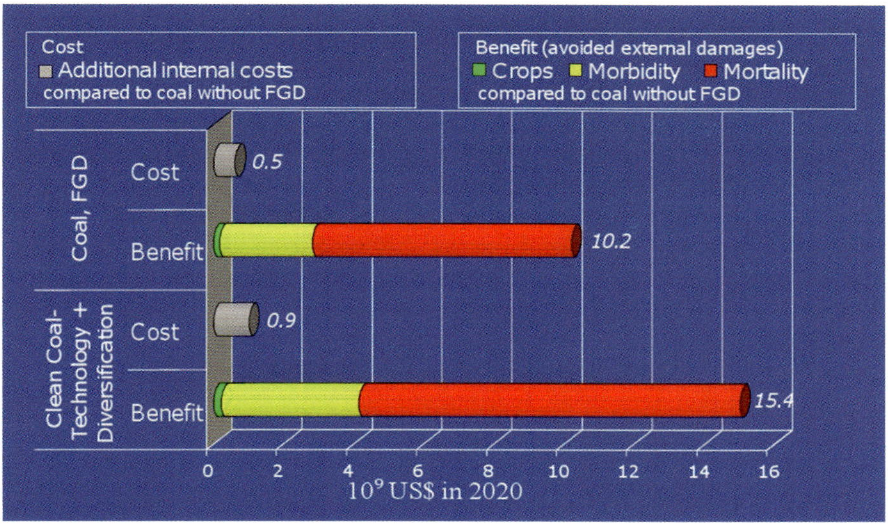

Fig. 8 Cost–benefit analysis of selected electricity generation technologies, scenarios for Shandong province, 2020 (Source: Hirschberg et al. 2003; Note: *FGD* flue-gas desulfurization, *incl. LCA* including life cycle assessment)

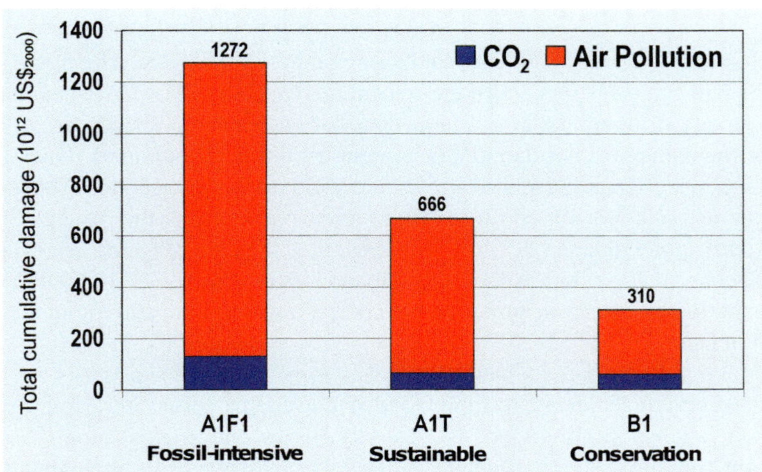

Fig. 9 Total cumulative damage due to increased CO_2 and air pollution under three IPCC scenarios, 1990–2100 (Source: Hirschberg and Burgherr 2002)

(Hirschberg and Burgherr 2002). Here, the IPCC emission scenarios (IPCC 2000) were used, i.e. three families of scenarios representing three levels of CO_2 emissions: the fossil-intensive scenario with the highest CO_2 emissions, the sustainable scenario with a lot of renewables, and the conservation scenario with enormous

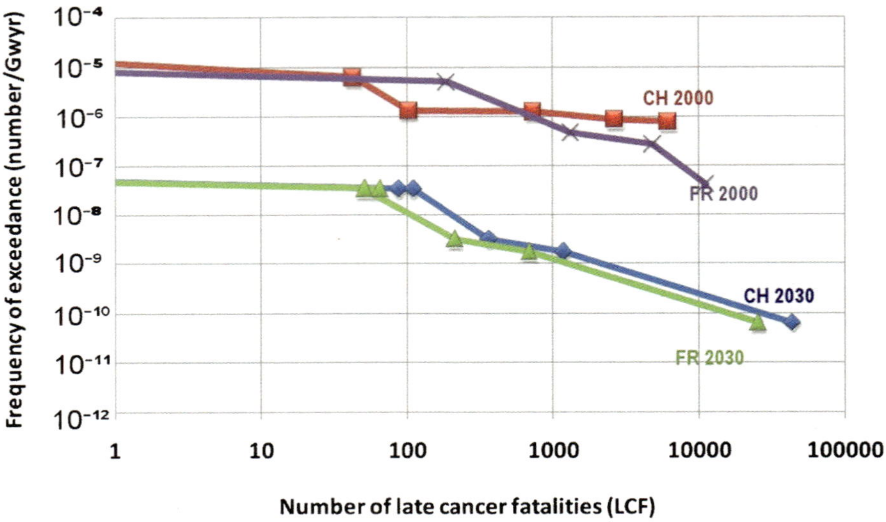

Fig. 10 Frequency–consequence (F–N) curves for nuclear energy: late cancer fatalities in Switzerland (CH) and France (FR) (Source: Hirschberg et al. 2008a; Note: FN curves show the relationship between the cumulative frequency (F) and the number of fatalities (N); *CH* Switzerland, *FR* France)

improvements in energy efficiency and also changes in lifestyles. The figure shows the cumulative damage for each of these scenarios (in US$ trillion). All the scenarios, although very different, include a large share of nuclear. A surprising result is that most of the damage is not due to CO_2 emissions, but to air pollution. Thus, implementing climate-friendly technologies could yield enormous secondary benefits by reducing air pollution. Based on the fossil-intensive scenario, the damage in 2100 would correspond to about 5.5% of global GDP. In the case of China, the estimated external costs of air pollution correspond to the loss of about 6–7% of GDP. In other words, the losses due to air pollution are comparable with the current annual GDP growth in China.

Estimating external costs inevitably raises some controversial issues, including the risk of severe accidents, terrorism and risk aversion, visual intrusion, resource depletion, nuclear proliferation, biodiversity losses and energy supply security. These are complex issues that determine social acceptance of technologies, related to social justice and to energy systems as the drivers of potential conflicts in society.

Considering these effects in terms of the corresponding costs tends to lead to low estimates, but the low estimates do not resolve the controversy. Figure 10 shows the estimates based on a probabilistic safety assessment (PSA) of frequency–consequence curves of late cancer fatalities due to a nuclear accident. The results illustrate the dilemma (Hirschberg et al. 2008a): while the estimated frequency of

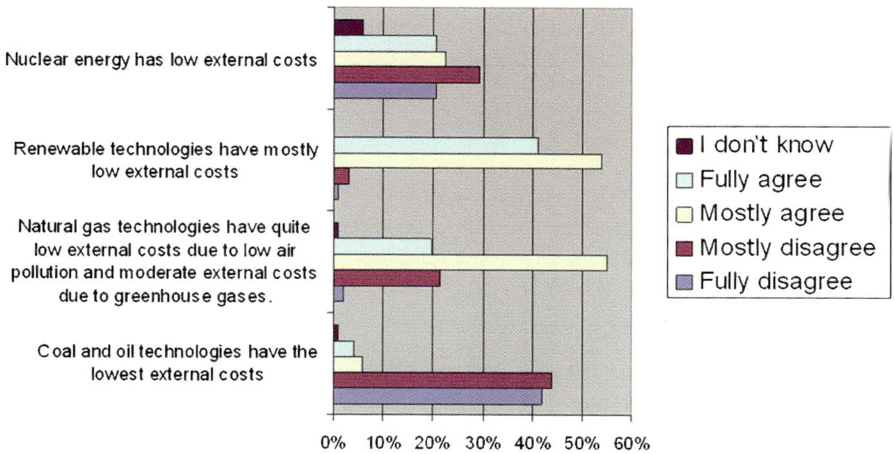

Fig. 11 Results of the NEEDS survey on the external costs of energy technologies (Source: Faberi et al. 2007)

hypothetical nuclear accidents with large consequences is very low, the possibility of very large consequences nevertheless exists. It is for this reason that many members of the public are opposed to nuclear power.

Figure 11 shows the results of a survey carried out during the NEEDS project to assess public acceptance of externalities associated with energy technologies (Faberi et al. 2007). The finding demonstrated that even if the above-mentioned limitations of the external cost framework are recognised, there is general acceptance of the concept of externalities, of the internalisation of external costs and of most of the results. A vast majority of stakeholders agree that most renewables have low external costs, that natural gas has low external costs due to air pollution but moderate external costs due to CO_2 emissions, and that oil and coal have high external costs. However, about 50% of stakeholders agree that nuclear has low external costs (as estimated in a series of EU-funded projects) while the other 50% disagree. This result is more a reflection of what people think about nuclear; those who are pro-nuclear easily agree that nuclear has low external costs, while those who oppose it believe the opposite.

6 A Complementary Approach: Multi-Criteria Decision Analysis

There is a complementary approach employing multi-criteria decision analysis (MCDA), based on a number of criteria. Table 3 lists the evaluation criteria developed by the NEEDS project (Hirschberg et al. 2008b), based on the three dimensions of

Table 3 Sustainability criteria

Dimension	Criteria
Environmental	*Resources*
	Energy resources
	Mineral resources (ores)
	Climate change
	Impacts on ecosystems
	Impacts from normal operations
	Impacts from severe accidents
	Wastes
	Special chemical wastes stored in underground depositories
	Medium- and high-level radioactive wastes to be stored in geological repositories
Economic	*Impacts on customers*
	Price of electricity
	Impacts on the overall economy
	Employment
	Autonomy of electricity generation
	Impacts on utility
	Financial risks
	Operation
Social	*Security/reliability of energy provision*
	Political threats to continuity of energy services
	Flexibility and adaptation
	Political stability and legitimacy
	Potential of conflicts induced by energy systems
	Necessity of participatory decision-making processes
	Social and individual risks
	Expert-based risk estimates for normal operations
	Expert-based risk estimates for accidents
	Perceived risks
	Terrorist threat
	Quality of residential environment
	Effects on the quality of landscape
	Noise exposure

Source: Hirschberg et al. (2008b)

sustainability: environmental, economic and social. The set of criteria was subjected to a European survey and found a very high level of acceptance, about 90%, by different groups of stakeholders (Hirschberg et al. 2008b). The difference between multi-criteria analysis and the total cost approach lies in their scope. Whereas the total cost method addresses components of economic and environmental dimensions and also health effects, multi-criteria analysis introduces a number of social issues that are important for the acceptance of certain technologies. MCDA involves the use of a broad set of criteria represented by quantitative indicators, combined

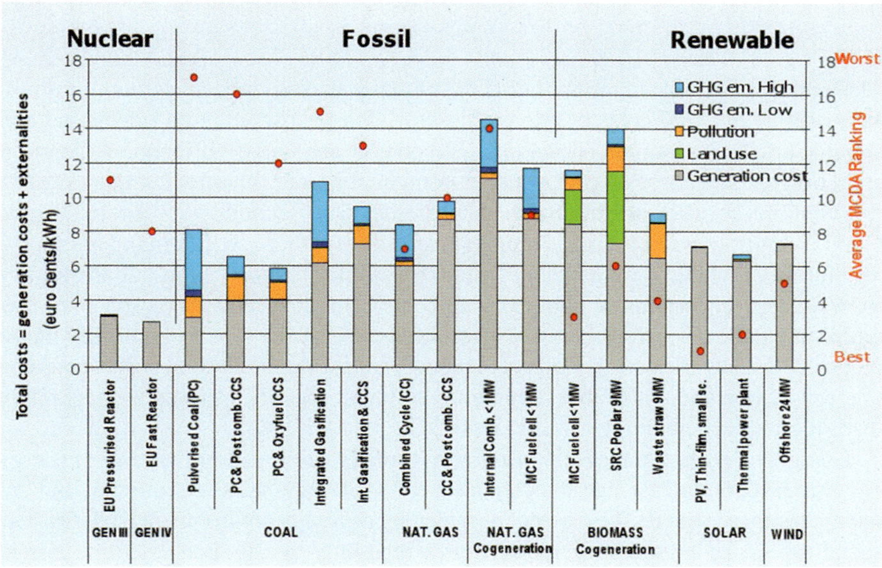

Fig. 12 Comparison of technology rankings based on total costs and those obtained using multi-criteria decision analysis (*red dots*) (Source: Schenler et al. 2009; Notes: *GHG em* greenhouse gas emissions, *GEN* generation, *NAT GAS* natural gas, *Post comb. CCS* post-combustion CO_2 capture and storage, *MC* molten carbonate, *SRC* short rotation coppice, *PV* photovoltaic, *MCDA* multi-criteria decision analysis)

with stakeholder preferences. Thus, different stakeholder groups have to make trade-offs and weigh the different criteria against each other.

In the NEEDS project, 36 indicators associated with the criteria listed in Table 3 were quantified using state-of-the-art methods. Stakeholders then had the opportunity to use an interactive web-based multi-criteria decision analysis tool developed by IIASA in cooperation with PSI. After indicating their preferences, stakeholders could then examine the resulting technology ranking and iterate their preference profile if they so wished. The evaluated technologies are highly advanced as the time horizon in the NEEDS project extends to 2050. Figure 12 compares the rankings based on total costs with the overall results of MCDA (Schenler et al. 2009). It shows that nuclear has the best ranking based on total costs. But based on MCDA nuclear has a lower ranking in the midfield, primarily because a wider set of indicators, especially social ones, are considered. The fossil-based systems mostly have the worst rankings with MCDA, except for natural gas combined cycle plants, whose ranking is comparable with that of nuclear. Renewables, particularly solar technologies, which do not have the best total costs, are likely to benefit from technological advances in the future, and most have top ranking with MCDA.

7 Conclusions

The health effects due to air pollution and global warming impacts dominate current estimates of external costs of energy systems. These costs may be substantial and should affect investment decisions, but there are large differences between technologies and locations. Consistent comparisons of technologies must clearly differentiate between technologies of yesterday and technologies for tomorrow. Good technologies, including advanced fossil, have rather low external costs due to air pollution, while climate change contributions may be very high but are subject to very large uncertainties. The external costs of nuclear and renewables are low within the estimation framework. Nevertheless, the ranking of technologies based on external costs is relatively robust. The internalisation of external costs should be vigorously pursued, since this is economically and socially justified and definitely leads to more efficient overall economies.

Using total costs as a measure of economic and environmental efficiency of energy systems favours nuclear, but the estimates for nuclear remain controversial and are only partially accepted. They are not accepted by the major environmental NGOs. The social aspects of energy systems are represented only to a limited extent in current estimates of external and total costs. Finally, the inclusion of broader social factors in the multi-criteria decision analysis framework favours renewables and is a challenge for future nuclear because it will have to contend with issues such as hypothetical large-scale accidents, the long-term confinement of waste and nuclear proliferation. Still, the total cost assessment remains a very useful tool for cost–benefit analysis.

References

Currie, J., & Walker, R. (2009). *Traffic congestion and infant health: Evidence from E-Zpass* (NBER Working Paper No. 15413). Cambridge: National Bureau of Economic Research. www.nber.org/papers/w15413

Desaigues, B., Ami, D., Hutchinson, M., Rabl, A., Chilton, S., et al. (2006). *Final report on the monetary evaluation of mortality and morbidity risks from air pollution* (Deliverable no. 6.7–RS1b), New Energy Externalities Development for Sustainability (NEEDS) project, Brussels, Belgium. www.needs-project.org

DLR, EDF, PSI, IER, POLITO, et al. (Deutsches Zentrum für Luft- und Raumfahrt, Institute of Technical Thermodynamics, Electricité de France, Paul Scherrer Institute, Institute of Energy Economics and Rational Use of Energy, Politecnico di Torino et al.). (2009). *External costs from emerging electricity generation technologies.* Deliverable no. 6.1–RS1a, New Energy Externalities Development for Sustainability (NEEDS) project, Brussels, Belgium. www.needs-project.org/2009

European Commission. (1999). *ExternE – externalities of energy.* Final Report. www.externe.info

Faberi, S., Gaggi, S., & Hirschberg, S. (2007). *Reporting and evaluation of the survey on the 'Externality' concept* (Deliverable report RS2b D-12.2), New Energy Externalities Development for Sustainability (NEEDS) project, Brussels, Belgium. www.needs-project.org

Friedrich, R. (2007). *The externE approach to assess marginal external costs of energy conversion.* Presentation at a NEEDS Workshop, Ljubljana, Slovenia. www.needs-project.org/public/Friedrich.ppt.zip

Hirschberg, S., & Burgherr, P. (2002). *Risks and benefits of climate-friendly energy supply options.* Invited paper presented at the sixth international conference on Probabilistic Safety Assessment and Management (PSAM6), San Juan, Puerto Rico.
Hirschberg, S., Heck, T., Gantner, U., Lu, Y., Spadaro, J. V., et al. (2003). Environmental impact and external cost assessment. In B. Eliasson & Y. Y. Lee (Eds.), *Integrated assessment of sustainable energy systems in china: The China energy technology program* (Alliance for global sustainability series, Vol. 4, pp. 445–586). Dordrecht: Kluwer.
Hirschberg, S., Bauer, C., Burgherr, P., Cazzoli, E., Dones, R., et al. (2008a). *Treatment of risks in sustainability assessment of energy systems.* Paper presented at the 9th international conference on Probabilistic Safety Assessment and Management (PSAM9), Hong Kong, China.
Hirschberg, S., Bauer, C., Burgherr, P., Dones, R., Schenler, W., et al. (2008b). *Sustainability criteria and indicators for assessment of electricity supply options* (Deliverable report RS2b D-3.2), New Energy Externalities Developments for Sustainability (NEEDS) project, Brussels, Belgium. www.needs-project.org/2009
Holmes, N. S., & Morawska, L. (2006). A review of dispersion modelling and its application to the dispersion of particles: An overview of different dispersion models available. *Atmospheric Environment, 40*(30), 5902–5928.
IPCC (Intergovernmental Panel on Climate Change). (2000). *Special report on emission scenarios* (Special report of Working Group III of the IPCC). Cambridge: Cambridge University Press.
Jamal, H. H., Pillay, M. S., Zailina, H., Shamsul, B. S., Sinha, K., et al. (2004). *A study of health impact & risk assessment of urban air pollution in Klang Valley, Malaysia.* Kuala Lumpur: UKM Pakarunding Sdn Bhd.
Karimzadegan, H., Rahmatian, M., Farhud, D. D., & Yunesian, M. (2008). Economic valuation of air pollution health impacts in the Tehran area, Iran. *Iranian Journal of Public Health, 37*(1), 20–30.
Marshall, J. D., Nethery, E., & Brauer, M. (2008). Within-urban variability in ambient air pollution: Comparison of estimation methods. *Atmospheric Environment, 42*(6), 1359–1369.
Mayer, C., & Sinai, T. (2003). Network effects, congestion externalities, and air traffic delays: Or why not all delays are evil. *American Economic Review, 93*(4), 1194–1215.
OECD (Organisation for Economic Co-operation and Development). (2001). *Extended producer responsibility: A guidance manual for governments.* Paris: OECD.
OECD (Organisation for Economic Co-operation and Development). (2005). *Analytical framework for evaluating the costs and benefits of extended producer responsibility programmes.* Paris: OECD.
Parry, I. W. H., Walls, M., & Harrington, W. (2007). Automobile externalities and policies. *Journal of Economic Literature, 45*(2), 373–399.
Richardson, H. W., & Bae, C.-H. C. (Eds.). (2008). *Road congestion pricing in Europe: Implications for the United States.* Cheltenham: Edward Elgar.
Roth, S., Hirschberg, S., Bauer, C., Burgherr, P., Dones, R., et al. (2009). Sustainability of electricity supply technology portfolio. *Annals of Nuclear Energy, 36*, 409–416.
Schenler, W., Hirschberg, S., Burgherr, P., & Makowski, M. (2009). *Final report on the sustainability assessment of electricity supply options* (Deliverable no. D10.2 – RS2b), New Energy Externalities Development for Sustainability (NEEDS) project, Brussels, Belgium. www.needs-project.org/2009
Sen, A. K., Tiwari, G., & Upadhyay, V. (2010). Estimating marginal external costs of transport in Delhi. *Transport Policy, 17*(1), 27–37.
Stein, A. F., Isakov, V., Godowitch, J., & Draxler, R. R. (2007). A hybrid modeling approach to resolve pollutant concentrations in an urban area. *Atmospheric Environment, 41*(40), 9410–9426.
Steyn, D. G., & Rao, S. T. (2010). *Air pollution modelling and its application XX* (NATO science for peace series C: Environmental security). Dordrecht: Springer.
Tol, R. S. J. (2006). *The polluter pays principle and cost–benefit analysis of climate change: An application of FUND* (Working Paper 2006.88). Milan: Fondazione Eni Enrico Mattei.
UN WWAP (United Nations World Water Assessment Programme). (2003). *Water for people, Water for life* (1st UN World Water Development Report). Paris: UNESCO and Oxford: Berghahn Books.

UN WWAP (United Nations World Water Assessment Programme). (2006). *Water, a shared responsibility* (2nd UN World Water Development Report). Paris: UNESCO and Oxford: Berghahn Books.

UNEP (United Nations Environment Programme). (2010). *Clearing the waters: A focus on water quality solutions*. Nairobi: UNEP.

WHO (World Health Organization). (2010). *Health and environment in Europe: Progress assessment*. Copenhagen: WHO Regional Office for Europe.

Part III
Energy Supply

Technology and Innovation

John H. Gibbons

Abstract The twentieth century has witnessed an extraordinary progression of technologies for both energy sources and energy use. This chapter reviews three 'enablers of change' both of supply and demand for energy. The first is technological invention and innovation, including basic research, exploration and mining technologies, oil and gas discovery and coal mining. The second enabler is the learning curve, an empirical function depicting declining costs of products with accumulating experience, for example, for renewable energy. The third enabler is the declining marginal costs of production as production units become bigger, but saturation can be observed as the negative externalities of large-scale production overwhelm the capacity of the natural environmental system to absorb unwanted emissions. This is particularly worrying in the case of global climate change and its implications.

1 Introduction

This chapter focuses on *energy services* rather than *energy supply* because energy services do not distinguish between the supply and the demand. We make our choice between adding supply or modifying demand based on relative total costs. While they may seem to be different universes, they actually must be considered part of the same universe of using energy services to help human needs and aspirations.

We have seen in the twentieth century an extraordinary progression of technologies for both energy sources and energy use. There has been an extraordinary revolution in services that basically replace human labour with energy. We have

J.H. Gibbons (✉)
Division on Engineering and Physical Sciences, National Research Council, The National Academies, 500 Fifth St., NW, 20001 Washington, DC, USA

F.L. Toth (ed.), *Energy for Development: Resources, Technologies, Environment*, Environment & Policy 54, DOI 10.1007/978-94-007-4162-1_11,

seen a bewildering array of amenities: light, power, transportation, communication, medical diagnostics and treatment, heating, ventilation and air conditioning, chemical processing and waste management. There is a long list of amenities that have been provided through energy services. There seems to be a never-ending set of opportunities, and we can witness that from the number of new products for consumers that appear daily, all of which have the common requirement of an energy driver.

It reminds me of what Thomas Edison said when he set up the first delivery system of electricity in New York more than 100 years ago. He said, 'I don't sell *electricity*, I sell *energy services'*. I think we need to distinguish between those two terms, the more fundamental being that it is the *use* we make of energy that really concerns and interests us. 'Energy Expo', the World's Fair on energy back in the 1970s, was entitled 'Energy Turns the World'. I think that title is still relevant.

2 Enablers

It is important to reflect on the enablers of change, both of supply and demand for energy, the so-called 'opportunities'. I think of three. One is technological invention and innovation. It includes basic research, exploration and mining technologies, and oil and gas discovery. Unfortunately the age of super-giant fields is past, but there are still many other frontiers of supply. An example is coal mining. One hundred years ago it was a matter of perhaps a steam shovel (but even more so a mule and a wagon) to dig coal out of the ground to convert to useful forms of energy. That process has been massively transformed, and as a consequence coal has become very inexpensive in the way it is delivered. It is, however, very expensive in terms of the remaining costs when we mine coal with a large bucket that can carry as much as 500 tonnes of coal each time it is lifted up and moved.

The transformation and lower costs of coal have been enormously affected by the advance of technology of the mining process itself. In Appalachia in the eastern United States, there are about 300 mountain peaks, named many years ago – named peaks that have now disappeared because we are tearing down the mountains in order to get to the coal seams that lie below the surface. 'Mountain-top removal' is a way to avoid the increased 'cost' of mining, but this coal extraction comes at a significant 'external' price. The real *cost* of that kind of extraction is not met by the selling price itself, and so we have major arguments now before the US Congress and other places around the world about how to account for the negative costs of coal mining. This is an example of the broader question of *cost versus price* that we must take account of in our energy systems.

What are the enablers of energy on both the supply and the demand sides? I do not distinguish between these two perspectives, supply and demand. They are largely interchangeable terms given the economic choices and the status of technology. First, enablers of change have to do with technological inventions and innovations that are derived from basic research, from exploration, from mining technologies, from oil and gas discovery mechanisms and from coal mining procedures. Resources that were out of reach until recently are now within reach, but at a higher cost. Advances in materials technology enable machinery to operate at higher temperatures and thus at higher efficiency.

The use of microwaves to heat our food, unheard of 50 years ago, is now a commonality in every kitchen of an electrified household. High-frequency compact fluorescent lamps provide four times as much light as incandescent lamps for the same energy consumed. The emerging market of light-emitting diodes will more than double the efficiency of fluorescent lighting. Advances have been made in battery technology, including nanotechnology, which in short is a way of dividing particles to an ultimate degree of fineness. It means that the surface to volume ratio is enormously higher than for conventional small particles, and a higher surface to volume ratio means breakthroughs in capabilities of electricity storage and other battery-like functions (Serrano et al. 2009; Garcia-Martinez 2010; Korkin et al. 2010). This is all relatively new but is already being widely applied not only to solar cells but to batteries themselves. So technological invention and innovation are very important enablers of change and improvements both in the supply of energy and in the demand for energy.

The second enabler or 'opportunity' is labelled a learning curve (van der Zwaan and Rablc 2004; Jamasb and Köhler 2007; Ferioli et al. 2009). Engineers and economists especially understand what a learning curve is. It is a very powerful empirical function that says basically that the more you make of something the less expensive the product will become. There is a falling cost curve that is inversely related to the log of the total production. It is almost a universal industrial theorem. The learning curve for energy is extraordinary, and we have seen it in the costs, for example, of renewable energy. We have also seen it in the combustion of coal and in other more traditional sources of supply. So let us remember the learning curve, because it applies both to existing technologies and particularly to new technologies.

The third enabler is the declining marginal costs of production (Maloney 2001; Rhine 2001; Iimi 2003; Blanco 2009). Declining marginal costs reflect the economy of scale. We have learned how to make bigger units, for example, multi-gigawatt power plants, and in turn the cost per unit of output has gone down. The problem is that while it is happening, we run into a saturation of that 'economy'. It soon becomes a 'diseconomy' of scale because the negative externalities of that size of production overwhelm the capacity of the natural environmental system to absorb unwanted emissions. The result is that, as size grows, economy of scale can transition to *diseconomy* of scale.

The lower cost of capital also has had a big impact, especially on the energy supply side. Large companies are able to obtain capital at a cost that is much lower than it used to be. Unfortunately, the individual consumer does not have that same advantage of low cost of capital. That shifts the economic decision toward the supplier rather than toward efficiency of use by consumer action.

3 Constraints

What about constraints on the growth of supply? There are negative externalities, and there are a number of ways to describe them. Let us consider the loss of high-grade resources. As previously noted, we now have to take the tops off mountains in order to extract coal, whereas 100 years ago it took a mule and a wheelbarrow. So we require greater investments in order to have that product.

Internationally, there are uneven and insecure distributions of resources, and that fact becomes more important all the time. An example is the Unites States, where we have used up high-grade fossil energy resources. We struggle with burgeoning negative environmental impacts of mining, processing, generation and use. Also, as the quantity of energy has been increasing all the time (for good cause as it comes with economic growth), the environmental impacts are more and more evident and pernicious (Rabl and Spadaro 2006; Bergmann et al. 2007; Tucker et al. 2008; Hirschberg 2012). These impacts are no longer absorbed by the capacity of Earth to absorb them without ill. These effects have been thought of in the old days as sort of local, maybe within a given state, maybe within a given country, but now we all understand that these effects are truly global, affecting the whole global climate (IPCC 2007). By climate I mean the temperature and level of our oceans and the extent to which we have flooding and droughts – a plethora of ill effects that come from overloading local environments and now are affecting the global environment.

The costs that are left behind are staggering, especially from global climate change, mainly caused by CO_2 emissions. That reality has been known about for probably 100 years or more, but our new understanding of the impact of human activities on the production of CO_2 is staggering to us. We still do not quite know how to accommodate this new reality. In fact there are many people, certainly many in the United States, who simply want to deny global climate change as being caused by human activities because it is so disturbing. It is so unnerving to think that we are affecting the whole globe with our energy-related activities. We tend to want to deny rather than to accept a less-than-one probability that we are moving the Earth into a situation that it has never endured before, or at least in the last few million years. Yogi Berra, a famous baseball catcher, once said, ‘If there is a 50–50 chance that something will go wrong, then nine out of ten times it will’. I think the probability of catastrophic climate change is something we must take seriously even if the probability is significantly less than one.

There is a saturation of economy of scale that now overtakes its improvement or its positive impacts – a *diseconomy of scale*. We have to take into account the diseconomies that we did not have to worry about 20 or 30 years ago. This occurs in all sorts of areas. For instance, a new idea on energy supply that could be a ubiquitous opportunity comes from new evaluations of the potential for geothermal energy extraction – not necessarily extremely high-temperature geothermal – because to heat a building or a house requires a relatively low-temperature geothermal source. For example, in upstate New York it is possible to install three geothermal wells several hundred feet down in the ground and to extract enough heat from the Earth itself to provide for both cooling and heating of an entire house that is located in a relatively cold climate. This is an example of an under-appreciated asset, namely the results of radioactive decay of heavy elements in the Earth's core.

In wind energy we have externalities and diseconomies (de Lucas et al. 2007; NRC 2007; Everaert and Stienen 2007; Tucker et al. 2008). We know that the best place on land to capture wind happens to be on high mountain ridges, but that is just where the birds of prey fly, and we have great morbidity and mortality of these birds when we try to put these wind turbines up on ridge lines. That is a significant diseconomy not only in terms of monetary costs but also in terms of the value as perceived by society of that kind of energy source. We therefore have to be somewhat more thoughtful and use special criteria in our siting of these turbines. There are a lot of constraints on the growth of supply. They are manageable, but they must be taken into account. It broadens the technical requirements of those who are involved in the siting and analysis of locations for power plants.

There are constraints also on energy efficiency. One constraint is market awareness. A brief example: in the Soviet Union after World War II they built a large number of residential structures. It turns out that the architects put the heating pipe systems on the outside of the buildings. They lost a lot of heat that way, but to make things worse they decided not to put thermostats on these systems but rather to let the residents open their windows when it got too warm inside in the wintertime. This is not exactly a conservative energy control system. Also, in the United States, there was a person, and this is a true story, in the Tennessee Valley who had a heating problem, and so she called the Tennessee Valley Authority (TVA) for assistance. She told them she had a big electric heater that she had put in her fireplace in the wintertime, but it simply could not heat her house. The TVA engineer went out, and indeed he found that there was this big electric heater sitting inside the fireplace. He looked up the chimney and saw there was no damper on the chimney. The electric heater was heating OK, but all the heat was going straight up the chimney. She did not understand that or appreciate the problem, and it was easily fixed. It is lack of awareness of cause and effect that is so easy to overcome and offset our so-called advantages.

There are constraints on efficiency of use that include not only market failures such as lack of consumer knowledge but also the typically higher 'first cost' of energy-consuming devices versus the purchase of power direct (Rutherford et al. 2007; Tonn and Peretz 2008; Dianshu et al. 2010). A consumer, unless he is very enlightened, will choose a cheaper first cost, which is usually the higher long-term cost. There is a lack of economic analysis of least total cost, 'least lifetime cost'. So it is an imperfect market place that distorts the rational allocation of funds between energy efficiency and energy supply.

4 Science and Technology Policy

For all economies, greater investment in and dependence on science and engineering are our greatest hope. In electricity production we need advanced fossil technologies such as carbon dioxide capture and storage (CCS) (IPCC 2005). Many know about

CCS: it does not work yet. There is no commercial unit in operation today, and yet we see advertisements all the time in the United States about ‘clean coal’. It is a wish, and that is as far as it goes for now until we get serious about the underlying science and technology of turning inherently dirty coal combustion into something that is clean. We can all too easily grasp at that appealing idea (clean coal), even though there are not yet cost-effective integrated systems that generate electricity and capture and store CO_2.

We also have a lack of appropriate policy tools to offset market failures on the demand side. Consumer awareness of what a most cost-effective choice would be is still very imperfect, even though we label efficiency. There is a recent study from McKinsey & Company (Granade et al. 2009) about the life cycle costing of various kinds of energy utilisation technologies. The study finds that in the United States today, if we made rational decisions about our energy investment, we could save US$1.3 trillion of energy supply by investing cost-effectively in energy demand efficiency. These are big numbers. For countries developing rapidly – that is, third world countries – it is even more imperative than for the mature industrial countries to make these investment decisions carefully, taking into account life-cycle costing. I am afraid we are not doing a very good job of that. I hope that our mistakes can be helpful to those who have not yet made those decisions.

5 Conclusions

I conclude this chapter by highlighting three points. First, time is not on our side in terms of global climate change and its implications for us. Time is not on our side because of facts of the matter that are very disturbing. A recent comprehensive analysis of climate change conducted at the Massachusetts Institute of Technology (Webster et al. 2009) shows that things are getting worse in terms of global effects of CO_2 enrichment, perhaps twice as fast as we thought only a half dozen years ago. Time is also not on our side because of political foot-dragging. It is not pleasant to think about our own actions as driving the global climate, and to think about the constraints that fact puts on what we once thought of as an ultimately exponentiating world and future. President John Kennedy said, ‘The supreme reality of our time is the vulnerability of our planet. It is up to us to understand that as well as we can and apply it to the way we help make decisions and devise options’.

Second, there are lots of opportunities to do something about energy services, both supply and utilisation – enormous opportunities. There is, however, no single magic bullet that we can count on. Faith alone can offer hope, but faith can also be an escape from reality, and that is an alluring illusion that many of us suffer from. As Mark Twain said, ‘I respect faith, but doubt is what gets you an education’. Doubt is the by-word of physics. At least, it is doubt that gives you a chance to learn things.

Finally, I note that we are all in this together. We certainly are roommates on Planet Earth whether we like it or not. We must develop and invest in human ingenuity, share in discovery and devise new measures and means of progress for mutual benefit. It is probably the first time that we have had such compelling common cause and purpose. I would urge everyone to remember two things. First, there is some seemingly unfathomable human potential, a true miracle – the extent to which humans apparently have the ability to envisage, to analyse and to act wisely if they choose to do so. Second, there are natural laws and an unbelievably energetic universe that follows those laws. 'God helps those who help themselves', said a wise person. But do not forget the admonition of Kenneth Boulding, the famous economist: 'The only people who believe that you can have infinite growth in a finite system are economists and other madmen'.

References

Bergmann, A. E., Colombo, S., & Hanley, N. (2007, April). *The social–environmental impacts of renewable energy expansion in Scotland*. Presented at the 81st annual conference of the Agricultural Economics Society, University of Reading, Reading, UK.

Blanco, M. I. (2009). The economics of wind energy. *Renewable and Sustainable Energy Reviews, 13*, 1372–1382.

de Lucas, M., Janss, G. F. E., & Ferrer, M. (2007). *Birds and wind farms*. Madrid: Quercus.

Dianshu, F., Sovacool, B. K., & Vu, K. M. (2010). The barriers to energy efficiency in China: Assessing household electricity savings and consumer behavior in Liaoning Province. *Energy Policy, 38*(2), 1202–1209.

Everaert, J., & Stienen, E. W. M. (2007). Impact of wind turbines on birds in Zeebrugge (Belgium). *Biodiversity and Conservation, 16*, 3345–3359.

Ferioli, F., Schoots, K., & van der Zwaan, B. C. C. (2009). Use and limitations of learning curves for energy technology policy: A component-learning hypothesis. *Energy Policy, 37*(7), 2525–2535.

Garcia-Martinez, J. (Ed.). (2010). *Nanotechnology for the energy challenge*. Weinheim: Wiley-VCH.

Granade, H. C., Creyts, J., Derkach, A., Farese, P., Nyquist, S., & Ostrowski, K. (2009). *Unlocking energy efficiency in the U.S. economy*. Zurich: McKinsey & Company/McKinsey Global Energy and Materials. McKinsey.com/clientservice/electricpowernaturalgas/dowloads/us_energy_efficiency_full_report.pdf. Accessed 4 June 2010.

Hirschberg, S. (2012). Externalities in the global energy system. In F. L. Toth (Ed.), *Energy for development: Resources, technologies, environment* (pp. 121–138). Dordrecht: Springer.

Iimi, A. (2003). *Economies of scale in power generation, transmission and distribution: Integration or unbundling?* (Working Paper No. 11), Tokyo: Japan Bank for International Cooperation.

IPCC (Intergovernmental Panel on Climate Change). (2005). *IPCC special report on carbon dioxide capture and storage* (Prepared by Working Group III of the IPCC). B. Metz, O. Davidson, H. C. de Coninck, M. Loos, & L. A. Meyer (Eds.). Cambridge: Cambridge University Press.

IPCC (Intergovernmental Panel on Climate Change). (2007). *Climate change 2007: The physical science basis* (Contribution of Working Group I to the fourth assessment report of the IPCC). S. Solomon, D. Qin, M. Manning, Z. Chen, & M. Marquis (Eds.). Cambridge: Cambridge University Press.

Jamasb, T., & Köhler, J. (2007). *Learning curves for energy technology: A critical assessment* (Cambridge working papers in economics). Cambridge: Cambridge University.

Korkin, A., Krstic, P. S., & Wells, J. C. (2010). *Nanotechnology for electronics, photonics, and renewable energy*. New York: Springer.

Maloney, M. T. (2001). Economies and diseconomies: Estimating electricity cost functions. *Review of Industrial Organization, 19*(2), 165–180.

NRC (National Research Council of the National Academies). (2007). *Environmental impacts of wind-energy projects*. Washington, DC: National Academies Press.

Rabl, A., & Spadaro, J. V. (2006). Environmental impacts and costs of energy. *Annals of the New York Academy of Sciences, 1076*, 516–526.

Rhine, R. (2001). Economies of scale and optimal capital in nuclear and fossil fuel electricity production. *Atlantic Economic Journal, 29*(2), 203–214.

Rutherford, J. P., Scharpf, E. W., & Carrington, C. G. (2007). Linking consumer energy efficiency with security of supply. *Energy Policy, 35*(5), 3025–3035.

Serrano, E., Guillermo, R., & Garcia-Martinez, J. (2009). Nanotechnology for sustainable energy. *Renewable and Sustainable Energy Reviews, 13*(9), 2373–2384.

Tonn, B. E., & Peretz, J. H. (2008). *Barriers to reducing energy consumption at home and on the road* (Working Paper), Knoxville: Institute for a Secure and Sustainable Environment/ University of Tennessee.

Tucker, G., Bassi, S., Anderson, J., Chiavari, J., Casper, K., & Fergusson, M. (2008). *Provision of evidence of the conservation impacts of energy production*. London: Institute for European Environmental Policy.

van der Zwaan, B., & Rablc, A. (2004). The learning potential of photovoltaics: Implications for energy policy. *Energy Policy, 32*, 1545–1554.

Webster, M., Sokolov, A. P., Reilly, J. M., Forest, C. E., Paltsev, S., et al. (2009). *Analysis of climate policy targets under uncertainty* (Report No. 180). Cambridge: MIT Joint Program on the Science and Policy of Global Change. www.globalchange.mit.edu/pubs/abstract.php?publication_id=1989. Accessed 4 June 2010.

Energy Resources

Hans-Holger Rogner

Abstract Energy resources are abundant. Their actual future availability in the market depends on a variety of constraints: production shifting to smaller deposits in increasingly harsh environments; rising exploration, production and marketing costs; excessive environmental burdens; diminishing energy ratios; and ever more stringent environmental policy and regulation. Energy demand, high prices and associated investments, innovation and technology change tend to increase flow rates to the market. Geological or other natural constraints will not impede the availability of energy resources for socioeconomic development. Mobilising resources might be hampered by inertia, long decision and investment cycles, as well as market uncertainties stemming from both the supply and demand sides. Forward-looking and stable policies are required to avoid abrupt supply and price fluctuations, and to mobilise the necessary investments for a socially optimal, long-term pathway for using energy resources that will foster sustainable development.

1 Introduction

Earlier chapters in this volume have demonstrated the importance of reliable and affordable energy supplies for socioeconomic development (Brew-Hammond 2012; Pachauri 2012; Luna and Gomelsky 2012). This chapter explores the fundamental

H.-H. Rogner (✉)
Department of Nuclear Energy, International Atomic Energy Agency (IAEA),
P.O. Box 100, Wagramerstrasse 5, A-1400, Vienna, Austria
e-mail: h.h.rogner@iaea.org

F.L. Toth (ed.), *Energy for Development: Resources, Technologies, Environment*,
Environment & Policy 54, DOI 10.1007/978-94-007-4162-1_12,

issues of energy supply: the global availability of energy resources over the long term. Since the steep oil price hikes of 2008, any discussion about the future availability of energy resources inevitably leads to a debate not only on ‘peak oil’ but also ‘peak gas, coal and uranium’. Some analysts use the peak resource argument to underpin their belief in the imminence of energy system transformation towards more rational energy use and renewable energy supplies. Indeed, in the case of oil, for many analysts the peak is here already (EWG 2007; ASPO 2011).

Clearly, for any finite resource that is continuously being mined or extracted, a point in time will be reached when production eventually peaks. The logic is straightforward. For example, oil is a finite resource that cannot be reproduced in human time frames. Any extraction further depletes the stock, and demand growth will rapidly lead to the exhaustion of remaining supplies. A finite stock plus rising demand equals depletion, increasing economic scarcity and peak production followed by inevitable decline.

Several tacit assumptions must be noted in this context. First, continuous production presupposes there is continuous demand for this finite resource. Second, the volume of the resource is known or defined a priori. Third, the geological occurrence and geophysical characteristics of the resource, e.g. concentration in a deposit or the technological mining conditions, are fully delineated. Fourth, technological changes in exploration and production technologies are not explicitly considered. Fifth, the costs of production have little or no impact on the marketability of the resource. Sixth, steeply rising market prices are an unmistakable sign of the imminent peak (the ‘depletion signal’).

Unlike members of the peak oil school – often labelled ‘peakists’ – some economists reject the notion of fixity (Zimmerman 1933). Rather, they view resources through the lens of demand or human needs. Resources are not an end in themselves but a means to satisfy human needs. Only the presence of demand turns the otherwise ‘neutral stuff’ in the Earth’s crust into valuable resources. Without demand, their physical availability is irrelevant. And today’s resources may not be tomorrow’s, and vice versa (Bradley 2004).

Resources, therefore, are not known, fixed things. Resource assessments were derided by Adelman (1992) as futile efforts to estimate the economic portion of an unknown and unknowable total. What matters is the timely availability of a resource in the market place at competitive costs. If a resource becomes too expensive the market responds in two ways: consumers tend to shift to alternative resources (demand reduction); and producers seek additional supplies through enhanced exploration activities and innovative production methods, thus enabling production from previously inaccessible deposits. During this transition period, resource availability is temporarily constrained and resource prices are highly volatile – a situation often interpreted as the sign of imminent resource scarcity.

Moreover, technology change and improvements in knowledge push the frontier of exploitable resources towards deeper, more remote or lower concentration occurrences, making resources a dynamically evolving rather than a ‘fixed’ quantity.

2 The Peak Resources Debate

The terms reserves, resources and occurrences are routinely used in the resource industry but there is no consensus on their exact meanings. Many countries and institutions have developed their own expressions and definitions, and different authors and institutions have different meanings for the same terms. This lack of consistent definitions and boundaries is one cause of confusion. Another is rooted in the fact that most resource quantities, estimated as deposits, are often located several kilometres below the surface. The estimates are based on inherently limited information and the geological data derived from exploration activities are subject to interpretation and judgment.

The dynamic nature of resources is well captured by the so-called McKelvey diagram (McKelvey 1967) shown in Fig. 1, which presents resource categories in a matrix that shows increasing degrees of geological assurance (knowledge) and economic feasibility. This scheme, developed by the US Bureau of Mines and the US Geological Survey (USGS 1980), is reflected in the international classification system used by the United Nations (UNECE 1997, 2010).

In this classification system, 'resources' are defined as 'concentrations of naturally occurring solid, liquid or gaseous material in or on the Earth's crust in such form that economic extraction is potentially feasible'. The geological dimension is divided into 'identified' and 'undiscovered' resources.

'Identified' resources are deposits whose location, grade, quality and quantity are known, or can be estimated from geological evidence. Identified resources are further subdivided into 'demonstrated' and 'inferred' resources, to reflect varying degrees of geological assurance, or lack thereof. 'Undiscovered' resources are

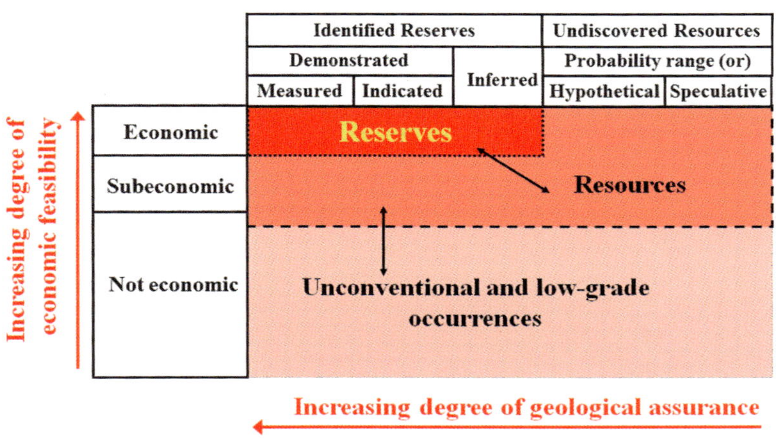

Fig. 1 Principles of resource classification (Source: Based on McKelvey 1967)

quantities expected or postulated to exist based on materials found in analogous geological conditions. 'Other occurrences' are materials that are too low grade or for other reasons not considered technically or economically extractable. For the most part, unconventional resources are included in this category.

The boundary between 'reserves' and 'resources' is the current or expected profitability of exploitation, governed by the ratio of future market prices to the long-run cost of production. Price increases and production cost reductions expand reserves by moving resources into the reserve category and vice versa. Production costs of reserves are usually supported by actual production experience and feasibility analyses, while cost estimates for resources are often inferred from current production experience, adjusted for specific geological and geographical conditions.

Reserve outlooks reported by the media are usually framed within a short-term market perspective, which focuses on prices, who is producing from which fields, where spare capacity might exist to meet short-term demand peaks, the politics of oil and how this all balances with demand.

The 'peak' debate focuses on the size of the reserves box of the McKelvey diagram in Fig. 1. Changing market prices for a mineral may expand or contract the economic dimension of the reserves box. Higher prices or price expectations allow the extraction of previously sub-economic deposits. Higher prices also stimulate advances in exploration and production technologies that lead to an expansion of both the knowledge and economic feasibility dimensions.

Declining prices tend to reduce the volume of reserves. The removal of coal mining subsidies in Germany caused a 90% drop in the country's reported hard coal reserves. This does not mean that the coal in the ground no longer exists (it now figures under resources), it just means that without subsidies the coal is too expensive to mine.

Reserve-to-production (R/P) ratios have been a popular indicator in the mineral industry signalling the years a resource would last based on current production. The ratio of currently known reserves over current production presents a static picture that ignores the dynamic development of demand, technological change and prices. For oil, the R/P ratio has fluctuated around 40 years for more than a century, while production has steadily increased (and the world should have run out of oil decades ago). The quasi-constant R/P ratio could only be the result of an equivalent increase in oil reserves. In terms of the McKelvey diagram, advances in the geosciences and production technologies shifted or converted previously uneconomic and/or unidentified oil occurrences from the resource domain to reserves.

Figure 2 shows the estimates of ultimately recoverable reserves (URR) of oil over time (oil is used here as a proxy for all finite resources). URR is an estimate of the total amount of recoverable oil that exists in the ground before first production starts. URRs, therefore, are typically broken down into three main categories: cumulative past production, reserves and resources. The term 'recoverable' makes URR a subjective estimate based on partial information that is subject to change – change that is at the root of the peak debate. Peak proponents consider URR to be fixed by geology and the laws of physics. Figure 2, however, shows that URR estimates tend to grow over an extended period of time. This results from advances in geosciences

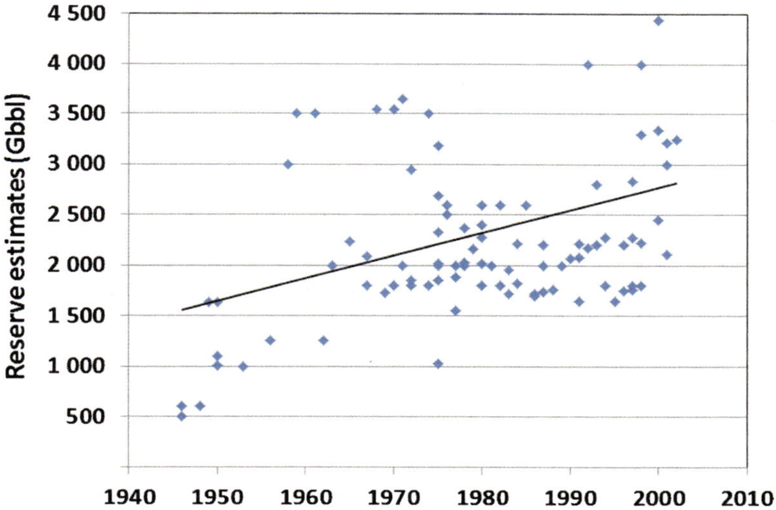

Fig. 2 Estimates of ultimately recoverable reserves (URR) of oil. Each point represents an independent estimate (Source: Adapted from Ahlbrandt and Klett 2005; NPC 2007; Note: *Gbbl* gigabarrel (10^9 barrel))

(knowledge), technology change (tapping locations previously beyond reach) and changed market settings (economics) – precisely the reasons why some economists often reject the fixed quantity URR concept. Rising URRs are consistent with the McKelvey diagram as they are the result of expanding dimensions of the 'reserves' box (geological assurance and economic feasibility).

The vast majority of recent URR estimates fall between 2,000 and 3,000 gigabarrels (Gb = 10^9 barrels) of oil (equivalent to 270 and 410 gigatonnes (Gt) of oil, respectively), of which about 1,150 Gb have been produced since the dawn of the oil age. For analysts supporting the lower URR range, therefore, the world has already produced 50% of URR and peak production is imminent (or already history), assuming continued growth in oil demand. Even the higher range would only postpone the peak by about two decades (depending on demand) before global oil production starts its inevitable decline.

The peak proponents argue for the lower URR range because reserve discoveries peaked between 1950 and 1960 and, as a consequence, the timing of peak production follows from the dynamics of past reserve additions. Moreover, for the last 30 years annual oil production has persistently exceeded annual reserve additions (see Fig. 3). Furthermore, the post-1980 reserve additions are predominantly virtual ones, i.e. not new discoveries but the result of field growth – the re-evaluation of previously discovered and already producing fields. The natural question then is why were the re-evaluated quantities not included in the original URR? Or are they possibly the result of improved production technology, better reservoir delineation or changed market settings as the economists maintain?

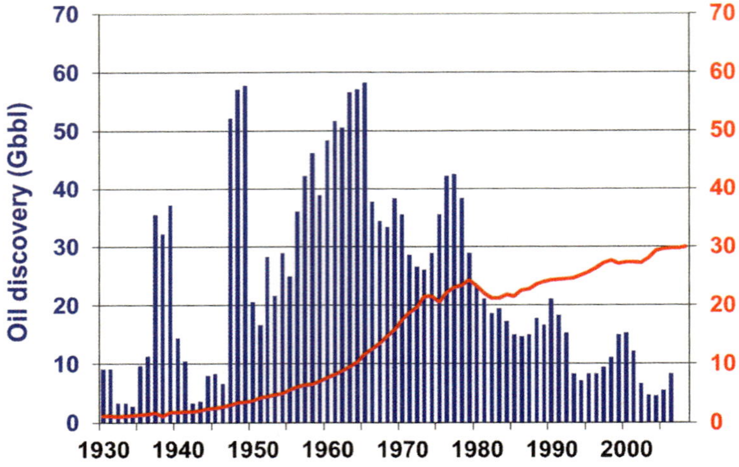

Fig. 3 Oil reserve additions (discoveries) and oil production (Source: Adapted from EPI 2007; Note: *Gbbl* gigabarrel (10^9 barrel))

In terms of market factors, higher prices not only push the frontier of marketable resources (exploitation of smaller fields, higher recovery rates, ability to operate in more challenging environments, etc.), but also stimulate upstream technology R&D for expanding exploration and production activities. At the same time, higher prices generally suppress demand, lowering pressure on supply. Claims that recent sky-rocketing oil prices reflect that the peak has arrived ignore these market fundamentals (as well as geopolitics, institutional priorities and environmental policy). As was observed in the second half of 2008, a slight reduction in global oil demand caused by the financial and economic crises started a precipitous drop in prices along the entire oil chain.

Using standard conventional production technologies based essentially on reservoir depressurisation, only around 20–40% of oil in place is recoverable by natural depletion of the reservoir. To extract further quantities, improved oil-recovery technologies, which can extract 30–60% of the oil in place, need to be applied (DOE 2010). The goal of enhanced oil recovery is to alter the original properties of the oil, to restore formation pressure, or to improve oil-displacement efficiency or fluid-flow patterns in the reservoir.

3 Conventional and Unconventional Resources

Another major factor in estimating future availabilities of exhaustible resources is the difference between 'conventional' and 'unconventional' occurrences (e.g. extra heavy oils, oil shale, tar sands, coal-bed methane, shale gas, methane clathrates or uranium dissolved in sea water). These terms lack standard definitions but are often used, which adds greatly to misunderstandings, especially in the debates on peak oil, gas or coal.

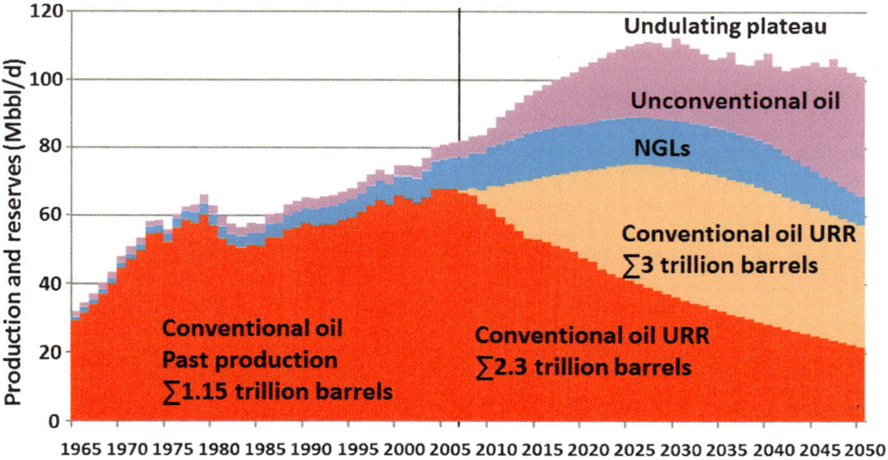

Fig. 4 Peak oil – which oil and when (Source: Adapted from CERA 2006; Note: *URR* ultimately recoverable reserve, *NGLs* natural gas liquids, *Mbbl/d* million barrel/day)

As the name suggests, unconventional resources generally cannot be extracted with technology and processes used for, say, conventional oil, which is usually understood as crude oil capable of flowing under normal conditions, i.e. oil with an API value greater than 20°. (API is an arbitrary scale designating an oil's specific gravity, a proxy for viscosity. The API scale equates high specific gravity values to low API gravity values, and vice versa.) Unconventional resources require different logistics and cost profiles, and pose different environmental challenges. Their future accessibility is, therefore, a question of technology development, i.e. the rate at which unconventional resources can be converted into conventional reserves (notwithstanding the demand for liquid fuels and relative costs). As new production methods are developed that allow unlocking unconventional resources at increasing rates, thus making them conventional mainstream products, the distinction between unconventional and conventional becomes meaningless. In short, like the boundary between conventional reserves and resources, the boundary between conventional and unconventional resources is in permanent flux.

Most important, 'unconventional' oil resources are not included in URR estimates until they become economically and technically producible, which was rarely the case at low oil market prices. The inclusion of unconventional oil resources in the standard future production profiles of the 'peak oil' analysts would radically change the shape of the global oil production profile (Odell 2004; McKenzie-Brown 2008). In fact, the notion of 'peak oil' is misleading. When total (conventional and unconventional) oil production approaches a maximum level, production is likely to be characterised by an 'undulating plateau' (see Fig. 4) rather than by a peak followed by a sharp drop-off in output (CERA 2006).

Figure 4 shows the past production profile of conventional oil of about one trillion barrels. Assuming the lower end of URRs estimated at two trillion barrels, the

Table 1 Fossil and uranium reserves, resources and occurrences[a] (Source: Rogner et al. forthcoming)

	Historical production through 2009	Production 2009	Reserves	Resources	Additional occurrences
Conventional oil (Gbbl)	1,164	29.2	860–1,333	730–1,080	
Unconventional oil (Gbbl)	106	4.0	660–980	1,970–2,600	>7,000
Conventional gas (Tcm)	93	3.0	134–190	193–240	
Unconventional gas (Tcm)	4	0.3	540–1,800	1,080–3,200	>25,000
Coal (Gtce)	248	5.2	590–720	9,930–14,800	
Conventional uranium[b] (ktU)	2,266	43.6	3,976	12,614	
Unconventional uranium (ktU)	n.a.			6,970	>4,000,000

Note: *Gbbl* gigabarrels (10^9), *Tcm* trillion cubic metres (10^{12} m^3), *Gtce* gigatonnes (10^9 t) of coal equivalent, *ktU* kilotonnes (10^3 t) of uranium (metal)

[a]The data reflect the ranges found in the literature; the distinction between reserves and resources is based on current (exploration and production) technology and market conditions

[b]Reserves, resources and occurrences of uranium are based on a once-through fuel cycle operation. Closed fuel cycles and breeding technology would increase the uranium resource dimension 50–60-fold. Thorium-based fuel cycles would enlarge the fissile-resource base further

world has indeed already reached the halfway point (or peak production) and future oil supply would essentially decline along the profile marked in red. Assuming the higher URR range of three trillion barrels (the salmon pink area in Fig. 4), then peak production would occur around 2030 and at a higher output level.

But conventional oil is not the only type of oil in the market place. Unconventional oil and natural gas liquids (NGLs) already supplement conventional oil supply in meeting rising global demand for liquid fuels (what markets demand and not oil per se). Unconventional oil resources are much more abundant than conventional oil reserves and resources (see Table 1). Their production is technically, economically and environmentally more challenging than conventional oil, but with growing demand expectations and sufficiently high market prices unconventional oil will be increasingly called upon to supplement conventional oil. In the past, human engineering ingenuity, technology change and innovation have countered upstream cost pressures and this is expected to continue. After all, 'knowledge is truly the mother of all resources' (Zimmerman 1951). Including unconventional oil resources and NGLs changes the potential global oil production profile as shown in Fig. 4.

Uncertainty as to the availability of unconventional oil is the timely investment in their production capacities as the oil produced would face the highest marginal costs, possibly an order of magnitude or more than the cheap conventional oil of the Middle East. In the recent past, investment in oil production capacity has not been a high national policy priority in several oil-exporting countries. Geopolitical market

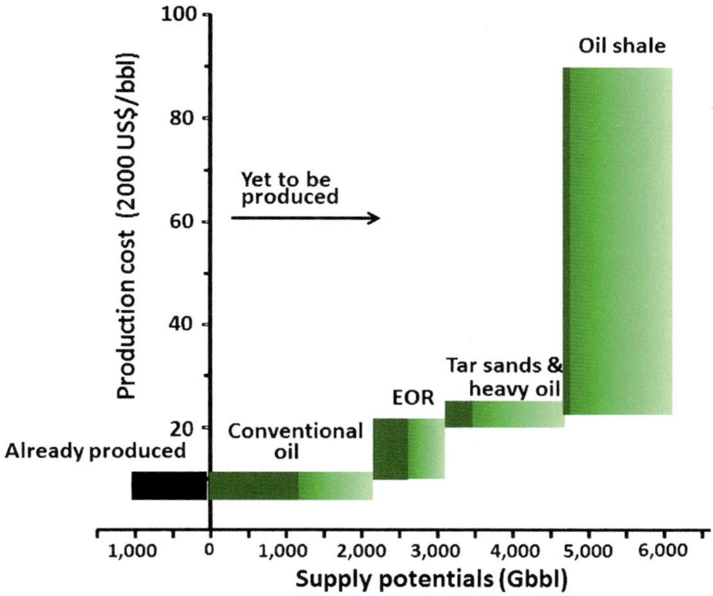

Fig. 5 Liquid fuel supply potentials and production costs (Source: Adapted from Farrell 2008; Note: *EOR* enhanced oil recovery, *Gbbl* gigabarrel (10^9 barrel))

manipulation in a fragile market situation (e.g. opening pumps in the short run) could flood the market with cheaper oil and put the investment in unconventional oil at risk.

Figure 5 plots a stylised long-term oil supply cost curve accounting for conventional and unconventional oil resources. The figure also puts the potential future oil production (right side of the vertical axis) in perspective with cumulative past production (left side of the vertical axis) and stacks the different oil categories according to their production costs. The height of each bar indicates the estimated production cost range, and the width indicates the prospective resource availability. The shading reflects the state of current knowledge about the resource – the darker shading points to a higher geological assurance, while the lighter shading indicate the more speculative nature of their existence and producibility.

The foregoing sections focused on oil because of its high visibility and its critical role for fuelling the global transportation system. But the same pattern holds for coal, natural gas or uranium. Table 1 summarises the overall energy resource outlook. Occurrences of hydrocarbons and fissile materials in the Earth's crust are plentiful. Yet despite numerous reviews, the range of resource estimates is large due to varying boundaries of what is included in the analysis of an exhaustible resource stock, e.g. conventional oil only or conventional oil plus unconventional occurrences, such as oil shale, tar sands and extra-heavy oils. Likewise, uranium resources are a function of the level of uranium ore concentrations in the source rocks that are considered technically and economically extractable over the long run, and, of course, demand.

Table 2 Renewable energy flows and resources (Source: Rogner et al. forthcoming)

	Utilisation 2008 (EJ)	Technical potential (EJ/year)	Annual flows (EJ/year)
Biomass	52.3	160–270	2,200
Geothermal	3.4	810–1,545	1,500
Hydro	12.8	50–60	160
Solar	1.0	62,000–280,000	3,900,000
Wind	2.5	1,250–2,250	110,000
Ocean	–	3,240–10,500	1,000,000

Note: 1 EJ = 10^{18} J

4 Renewable Resources

Renewable energy resources represent the annual energy flows available through sustainable harvesting on an indefinite basis. While their annual flows far exceed any conceivable global energy needs, the challenge lies in developing adequate technologies to harvest the often low or varying energy densities and overcome supply intermittencies, and to convert them into usable fuels.

Table 2 summarises the annual flows of renewable energy resources and compares them with their respective technical potentials – the degree of utilisation of renewable flows that is possible within thermodynamic limitations, geographical conditions (orientation, terrain, distances to demand centres) and, most important, the technological performance characteristics of the conversion technologies without considering their economic feasibility. The latter, however, determine the economic deployment potential of renewables (kind of the equivalent of reserves). Conversion technologies feeding on renewable energy flows, especially wind and solar, have experienced steady performance improvements and substantial market penetration, in large part as a result of public policy support, including subsidies. However, their overall contribution to global energy supply is unlikely to slow down significantly the demand for hydrocarbon resources in the near- to mid-term future.

5 Conclusions

Energy resources are abundant. But in Zimmerman's (1951) words: 'resources are not, they become'. The real question, therefore, concerns the actual future availability of energy resources in the market place. While the overall size of the 'barrel' is a matter of debate for geologists and irrelevant for others, what matters really are the sizes of the taps that enable and control the flow rates from the barrel to the market place. The taps are subject to a variety of constraints: production shifting to smaller and smaller deposits in harsher and harsher environments; rising exploration, production and marketing costs; excessive environmental burdens; diminishing energy ratios; and ever more stringent environmental policy and regulation. These factors can limit flow rates or rather the rate at which resources 'become'. In turn, energy demand, high prices (plus associated investments), innovation and technology change tend to increase 'become' and flow rates.

The question, then, is which of these opposing forces will prevail in the mid- to long term? Are the necessary above-ground investment decisions forthcoming in a timely fashion that will unlock below-ground resources? If that investment is not forthcoming, no matter how large the resource, it just will not 'become' and will remain 'neutral stuff'.

This chapter has demonstrated that geological or other natural constraints will not impede the availability of energy resources for fuelling socioeconomic development. However, mobilising the resources might be hampered by inertia, long decision and investment cycles as well as market uncertainties stemming from both the supply and demand sides. Forward-looking and stable policies are required to avoid abrupt supply and price fluctuations, and to mobilise the necessary investments for a socially optimal, long-term pathway for using energy resources that will foster sustainable development.

References

Adelman, M. A. (1992, June 23–24). *The international energy outlook*. Presented at International Energy Workshop, Harvard University, Boston.

Ahlbrandt, T., & Klett, T. (2005). Comparison of methods used to estimate conventional undiscovered petroleum resources: World examples. *Natural Resources Research, 14*(3), 187–210.

ASPO (Association for the Study of Peak Oil). (2011). www.peakoil.net

Bradley, R. L. (2004). *Are we running out of oil?* PERC Reports 22(3). Bozeman: Property and Environment Research Center. www.perc.org/pdf/sept04.pdf. Accessed 5 Sept 2011.

Brew-Hammond, A. (2012). Energy: The missing Millennium Development Goal. In F. L. Toth (Ed.), *Energy for development: Resources, technologies, environment* (pp. 35–43). Dordrecht: Springer.

CERA (Cambridge Energy Research Associates). (2006). In P. Jackson (Ed.), *Why the "Peak Oil" theory falls down: Myths, legends, and the future of oil resources*. Cambridge: CERA.

DOE (Department of Energy). (2010). *Enhanced oil recovery/CO2 injection*. DOE Oil Recovery R&D Program. http://fossil.energy.gov/programs/oilgas/eor/. Accessed 28 Dec 2010.

EPI (Earth Policy Institute). (2007). *Is world oil production peaking*? www.earth-policy.org/Updates/2007/Update67_printable.htm

EWG (Energy Watch Group). (2007). *Crude oil: The supply outlook report to the energy watch group* (EWG-Series No. 3/2007). Ottobrunn: Ludwig-Bölkow-Systemtechnik GmbH.

Farrell, A. E. (2008). Energy notes. *News from the University of California Energy Institute, 6*, 3.

Luna, N., & Gomelsky, R. (2012). Development, infrastructure and energy: Exploring the linkages in Latin America. In F. L. Toth (Ed.), *Energy for development: Resources, technologies, environment* (pp. 95–112). Dordrecht: Springer.

McKelvey, V. E. (1967). Mineral resource estimates and public policy. *American Scientist, 60*, 32–40.

McKenzie-Brown, P. (2008). *Colin Campbell and the cracks of doom*. http://languageinstinct.blogspot.com/2008/03/colin-campbell-and-crack-of-doom.html. Accessed 17 Jan 2009.

NPC (National Petroleum Council). (2007). *Hard truths: Facing the hard truths about energy*. Washington, DC: NPC.

Odell, P. (2004). *Why carbon fuels will dominate the 21st century's global energy economy*. Brentwood: Multi-Science Publishing.

Pachauri, S. (2012). Demography, urbanisation and energy demand. In F. L. Toth (Ed.), *Energy for development: Resources, technologies, environment* (pp. 81–94). Dordrecht: Springer.

Rogner, H.-H., Aguilera, R. F., Archer, C. L., Bhattarcharya, S. C., Dusseault, M. B., et al. (forthcoming). Energy resources and potentials. *In Global energy assessment: Towards a sustainable future.* Laxenburg/Cambridge: IIASA/Cambridge University Press.

UNECE (United Nations Economic Commission for Europe). (1997). *United Nations international framework classification for reserves/resources: Solid fuels and mineral commodities.* Geneva: ECE.

UNECE (United Nations Economic Commission for Europe). (2010). *United Nations international framework classification for fossil energy and mineral reserves and resources 2009* (ECE Energy Series No. 39. Geneva: ECE. http://live.unece.org/fileadmin/DAM/energy/se/pdfs/UNFC/unfc2009/UNFC2009_ES39_e.pdf. Accessed 5 Sept 2011.

USGS (US Geological Survey). (1980). *Principles of a resource/reserve classification for minerals* (US Geological Survey Circular 831).

Zimmerman, E. W. (1933). *World resources and industries: A functional appraisal of the availability of agricultural and industrial resources*. New York: Harper & Brothers.

Zimmerman, E. W. (1951). *World resources and industries*. New York: Harper & Brothers.

Financing Power Sector Investments

Carmine Difiglio

Abstract Global energy demand is projected to increase by about 33% between 2010 and 2030. In non-OECD countries, electricity generation capacity is forecast to grow by 2,900 gigawatts (GW) by 2030, while the OECD countries will add only 1,600 GW. The related investments needed are estimated at US$7.8 trillion and US$5.9 trillion for non-OECD and OECD countries, respectively. Private capital could finance much of this investment but there are major obstacles to private investments in the power sector in developing countries. Reform efforts that move towards competitive markets can help address these risks, paving the way for the injection of much-needed private capital. Power sector reform can contribute to economic, as well as social, development. Successful power sector reform reduces government fiscal responsibilities, improves the reliability of supplies for existing customers, and provides affordable access to energy for the poor.

1 Introduction

Global energy demand is projected to increase by about 33% between 2010 and 2030 with continued current government policies. Energy demand is not growing equally throughout the world. Non-OECD countries will experience more growth in energy demand through 2030 than OECD countries. Non-OECD countries are projected to add 2,900 gigawatts (GW) of electricity generation capacity from 2007 to 2030, while OECD countries will add only 1,600 GW. To provide a sense of scale, the world's largest electricity-generating plant, the Three Gorges Dam in China, has a capacity of

C. Difiglio (✉)
Office of Policy and International Affairs, 1000 Independence Ave., S.W., Washington, DC 20585, USA

F.L. Toth (ed.), *Energy for Development: Resources, Technologies, Environment*, Environment & Policy 54, DOI 10.1007/978-94-007-4162-1_13,

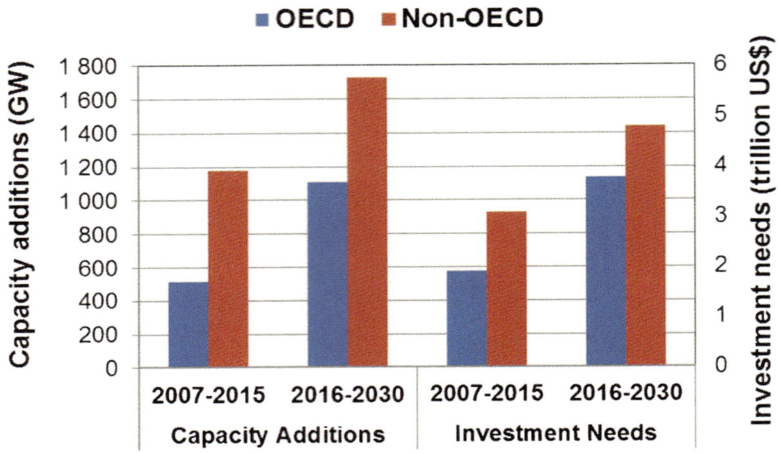

Fig. 1 Capacity additions and investment needs in OECD and non-OECD countries, 2007–2030 (Source: IEA 2008)

almost 20 GW, while the largest coal and nuclear power plants have capacities on the order of 5 GW. Large investments will be required to fund these projected capacity expansions, estimated at US$7.8 trillion for non-OECD countries and US$5.9 trillion for OECD countries between 2007 and 2030 (IEA 2008), as shown in Fig. 1.

Private capital could finance much of the needed investment. This is particularly important in developing countries where the public funds available for infrastructure development are limited. However, in order to attract private capital, investments must have a good chance of being profitable. Strong profitability has been a particular problem in the power sector in many developing countries.

This chapter identifies the main obstacles to private investment in developing countries, and approaches to overcome them. Section 2 describes the investment risks in energy infrastructure in developing countries. Section 3 identifies power sector reforms that could make the financing environment more inviting for private capital in non-OECD markets. Section 4 focuses on how successful implementation of these policies will lead to positive outcomes for investors, industry, governments and citizens. Finally, Sect. 5 presents a summary and conclusions.

2 Investment Risks in Developing Countries

2.1 Unstable Power Markets

Unstable or unreliable power markets represent a major risk to power sector investment in developing countries, exemplified by electricity shortages, technical losses and non-technical losses. Electricity shortages resulting in poor power quality and blackouts affect both consumers and producers. In South Africa, for example, regular blackouts have been experienced since 2008 (Bearak and Dugger 2008). Due to these blackouts, citizens have to deal with intermittent interruptions in all aspects of life, from slow travel due to traffic light outages, to shopping delays when cash registers shut down. Industrial users have had to accept rationed supplies of energy, resulting in reduced productivity and profitability. The South African economy has suffered accordingly, hindering future growth and future stability.

As well as their impacts on consumers, electricity shortages pose several challenges for power producers. First, it is more difficult to accommodate regular downtime for the maintenance of power plants. Yet without minimal maintenance, power plant lifetimes and thermal efficiencies may be reduced. Second, power shortages erode power quality, which increases the wear and tear on generators, and on transmission and distribution systems. Increased wear results in shorter lifetimes and increased repair costs, reducing the profitability of investments.

High technical losses from inadequate or outdated transmission and distribution systems reduce the profitability of power producers. Technical losses are those inherent to the electricity distribution system, where power dissipates in electric power components (power lines, transformers, etc.). They can be estimated based on the physical equipment used.

In developing countries, non-technical losses may exceed technical losses. Non-technical losses are those unrelated to the physical system, and include theft, poor metering of use and inadequate collection of bills. Together, technical and non-technical losses are referred to as commercial losses. Commercial losses have been measured at 30–40% in some developing countries, whereas they are closer to 10% in many developed countries. One study has estimated that a 1% reduction in commercial losses resulted in US$1 of additional revenue per customer for a medium-size utility in Latin America (Bakovic et al. 2003).

2.2 Government Policy Uncertainty and Regulation

Uncertainty about government policies and regulations also pose significant challenges to investment in the power sector in many developing countries. Investments in power generation, transmission and distribution are long-lasting, since the functional lifespans of these systems may span several decades. In many developing countries, it cannot be assumed that there will be a stable policy or regulatory regime. This is a particularly large problem if the government controls or operates industries or sectors of the economy.

Initiatives to develop an economy through industry, such as manufacturing, may rapidly increase demand for power. If these initiatives are highly uncertain beyond the short-term, it is difficult to raise capital for power sector investments. Revenues beyond the short term would have to be heavily discounted to reflect the uncertain environment. There may also be fears that governments may expropriate property, apply new taxes, establish price controls or introduce other regulations that would adversely affect anticipated revenues once investments have been made.

3 Power Sector Reform

Three key areas of power sector reform are economic tariff reforms, private management and competition. Economic tariff reforms, i.e. making consumers pay the costs of producing and supplying electricity, are intended to achieve a better economy-wide use of resources. Private management is important as the profit motive gives a stronger incentive for efficient use of inputs than any incentives offered by an enterprise controlled and managed by a bureaucracy. Competition, made possible by technical innovations that have allowed the unbundling of electricity supply chains, can be introduced in the market for generation and retail segments. It has also made competition possible for transmission and distribution segments, which are natural monopolies.

The standard model of power sector reform, proposed by the World Bank (1993), follows a series of steps towards the development of a competitive market (see Fig. 2). The first step of corporatisation and commercialisation involves separating the electric utility from the relevant ministry and obliging it to operate according to commercial principles. The second step involves developing economic regulation of the power market that is applied transparently by an autonomous regulator. Third, independent power producers (IPPs) can help the reform process by demonstrating the benefits of private investment and management. The fourth step is unbundling, or restructuring, the electric power supply chain to enable the introduction of competition. Fifth, the unbundled electricity segments are privatised under dispersed ownership. Finally, the development of competition in the generation and supply segments follows from the development of power exchanges.

Focusing on the first step towards creating competition in the power market, significant efficiency and accountability gains can be achieved by moving from a bureaucracy to professional management. Figures 3 and 4 present the general organisational structures of a bureaucracy and professional management, respectively. Professional management is a more horizontal organisational structure, with fewer middle managers and fewer levels of employees than in a bureaucracy. Professional management also establishes a clear division of tasks throughout the organisation. For example, several offices in a bureaucracy might be responsible for a set of tasks, whereas under professional management these responsibilities would be concentrated in a single office. A board of directors at the top of the organisational structure, coupled with a defined profit metric, provides greater accountability than a single minister of power with less clear and stable metrics.

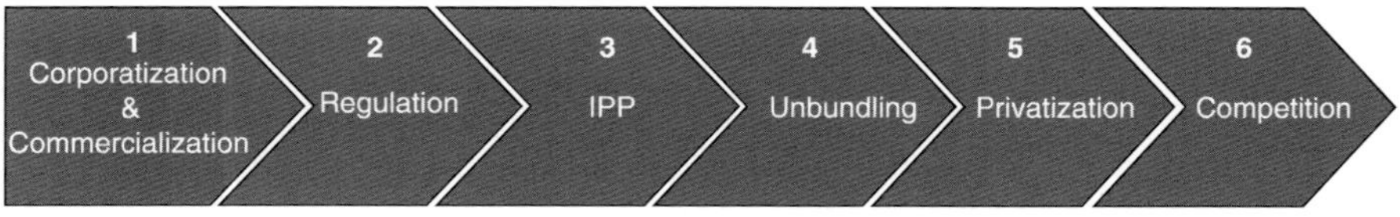

Fig. 2 Standard model of power sector reform (Source: World Bank 1993; Note: *IPP* independent power producer)

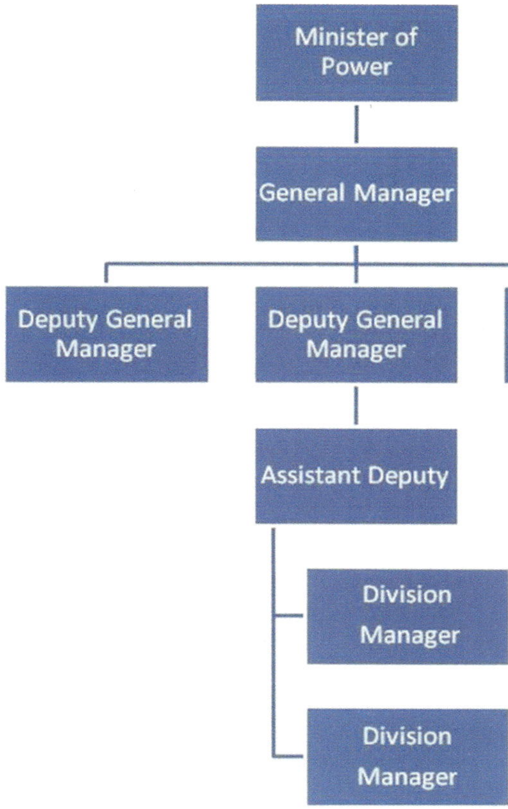

Fig. 3 General organisational structure of a bureaucracy

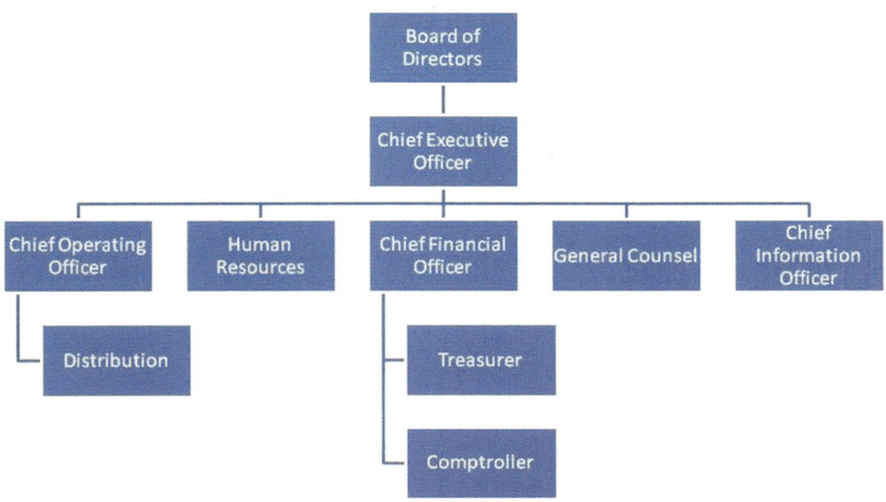

Fig. 4 General organisational structure of professional management

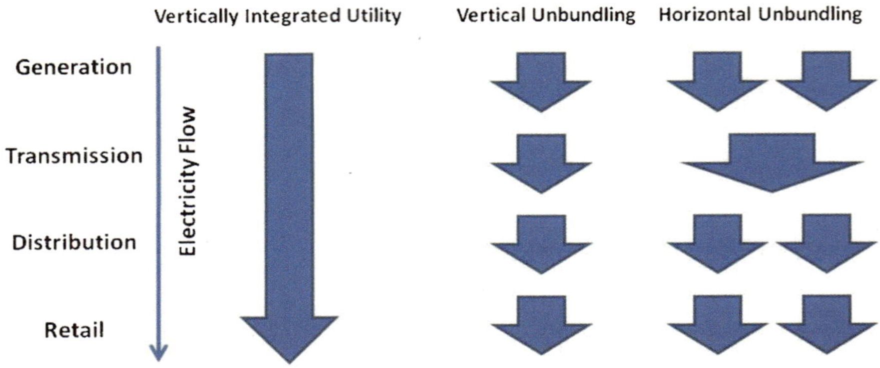

Fig. 5 Unbundling of vertically integrated utilities

The unbundling of vertically integrated utilities is an important step on the path towards competition. The power sectors in developing countries are generally characterised by vertically integrated utilities that combine generation (power production), transmission (high-voltage long-distance power lines and transformers), distribution (low-voltage local power lines) and retail sales to customers. Unbundling vertically integrated utilities into these four services isolates sectors of the industry where competitiveness is possible. Horizontal unbundling, i.e. increasing the number of players in each sector, directly increases competition in the sector (see Fig. 5).

Despite continued efforts over the past two decades, few developing countries have been able to adopt the standard model of reform completely. Bacon and Besant-Jones (2001) estimate that some countries have taken the key reform steps to varying degrees, as shown in Table 1. It is not clear whether these reforms merely require more time to implement than initially expected, or whether the standard model needs to be adjusted. One lesson that has been learned in the course of pursuing power sector reform is that the sequence of events may be more malleable than initially thought. For example, the introduction of IPPs does not necessarily have to follow other reforms. In some countries, IPPs have been introduced alongside vertically integrated utilities prior to the creation of regulators (Gratwick and Eberhard 2008). It may also be the case that the reform process would have been smoother had the suggested sequence of reforms been followed.

4 Expected Outcomes and Limitations

Electricity sector reforms in developing countries are meant to improve conditions for utilities, the government, electricity consumers and the poor. Utilities can expect better operational and financial performance. Governments will be relieved of excessive capital burdens, putting them in a better fiscal position. Electricity

Table 1 Number of countries that had taken key reform steps by 1998 (Source: Bacon and Besant-Jones 2001)

	Region (number of countries)					
Key step	Africa (48)	East Asia and Pacific (9)	Europe and Central Asia (27)	Latin America and the Caribbean (18)	Middle East and North Africa (8)	South Asia (5)
Corporatisation	15 (31%)	4 (44%)	17 (63%)	11 (61%)	2 (25%)	2 (40%)
Law	7 (15%)	3 (33%)	11 (41%)	14 (78%)	1 (13%)	2 (40%)
Regulator	4 (8%)	1 (11%)	11 (41%)	15 (83%)	0 (0%)	2 (40%)
Independent power producers (IPPs)	9 (19%)	7 (78%)	9 (33%)	15 (83%)	1 (13%)	5 (100%)
Restructuring	4 (8%)	4 (44%)	14 (52%)	13 (72%)	3 (38%)	2 (40%)
Generation privatisation	2 (4%)	2 (22%)	10 (37%)	7 (39%)	1 (13%)	2 (40%)
Distribution privatisation	2 (4%)	1 (11%)	8 (30%)	8 (44%)	1 (13%)	1 (20%)
Reform indicator[a]	0.88 (15%)	2.44 (41%)	2.70 (45%)	4.28 (71%)	1.00 (17%)	3.00 (50%)

[a]The 'Reform indicator' shows the average of the reform steps undertaken in the region

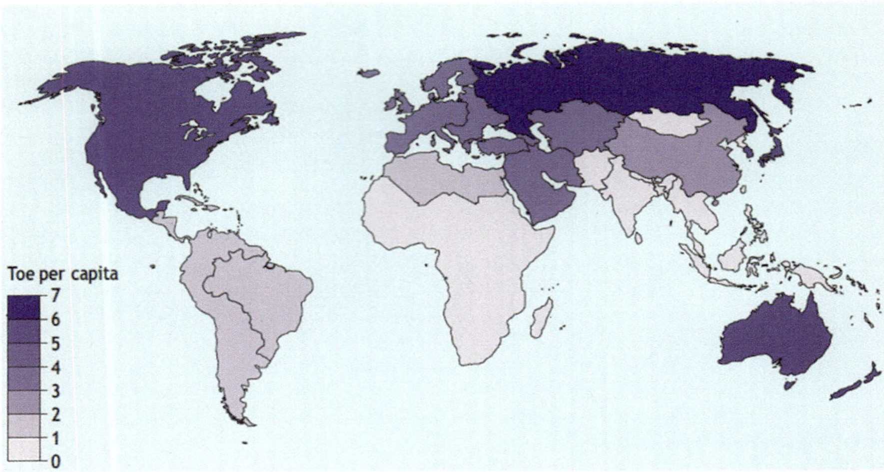

Fig. 6 Primary energy demand (tonnes of oil equivalent) per capita in 2030 (Source: IEA 2008)

consumers can expect more reliable services and improved quality power supplies. Many of the poor in developing countries will gain affordable access to electricity. Access to electricity provides many benefits, including improved safety, productivity and educational opportunities (Kanagawa and Nakata 2008; Gómez and Silveira 2010; Kooijman-van Dijk and Clancy 2010; Wilson et al. 2010).

There are several examples of improved fiscal conditions as result of power sector reform.

- In Bolivia, fiscal revenues from the power sector increased by 247% in 3 years. In addition, debt service of US$61 million, which was guaranteed by the government, was transferred to the private company (World Bank 1999).
- In El Salvador, the sale of 75% of shares in distribution companies ('discos') yielded US$575 million, which had a substantial financial impact equivalent to 5.5% of 1996 GDP.
- In Panama, in 2000, the privatised power sector companies contributed US$70.8 million to the treasury, of which US$34.5 million was in income taxes and US$36.3 in dividends for the shares still in government hands.
- In Peru, the sector shifted from being a drain on the public treasury of US$300 million in 1990, to being a source of fiscal income from US$300 million in profits in 1998 (Dammert et al. 2005).

Power sector reforms are not a panacea, however. Even if expansion of the power sector is fully funded, enabling the projected rate of growth, global energy consumption will continue to be inequitable in 2030. In some cases, developed countries will demand and consume seven times more primary energy per capita than other regions, as shown in Fig. 6. Further efforts will be required to support investment and improve energy access in developing countries in the future.

There are a host of issues that may remain despite successful implementation of power sector reform. For one, in a restructured electricity market, profit alone is often an insufficient driver for expanding access to electricity to relatively unprofitable rural customers and the urban poor. Incentive schemes, subsidies or regulatory mandates may be required. Electricity reforms are typically associated with pressures to limit subsidies and enhance the collection of tariffs. While these changes make for a better functioning sector, the resultant price increases can also cause social hardship and provoke political opposition to the reforms. Private companies that are unable to maintain viable commercial enterprises tend to pull out, leaving a deadlocked or confused situation in the power sector that may take years to sort out and may have a negative effect on other industries. In theory, other subsidies could be provided *in lieu* of subsidised electricity prices, although in practice this is not easily done. Removing or reducing price subsidies is politically difficult, and the databases and bureaucracies that are needed to administer alternative subsidy programmes may not be available. Such alternative subsidy mechanisms may also cause misallocation of resources in other areas.

5 Conclusions

To achieve even conservative estimates of the growth in power sectors in developing countries will require huge capital outlays. Private investment will be a critical contributor to this endeavour. Reform efforts that move towards competitive markets can help to address the investment risks in the power sectors of developing countries, paving the way for the injection of much-needed private capital.

Power sector reform can contribute to economic, as well as social, development. Successful power sector reform reduces government fiscal responsibilities, improves the reliability of supplies to existing customers, and provides affordable access to the poor. There may be additional stumbling blocks even with successful power sector reform. Nonetheless, power sector reform is an important step in the right direction in developing countries.

References

Bacon, R. W., & Besant-Jones, J. (2001). Global electric power reform, privatization, and liberalization of the electric power industry in developing countries. *Annual Review of Energy and the Environment, 26*, 331–359.

Bakovic, T., Tenenbaum, B., & Woolf, F. (2003). *Regulation by contract: A new way to privatize electricity distribution* (Working Paper 14). Washington, DC: World Bank.

Bearak, B., & Dugger, C. W. (2008, January 31). Power failures outrage South Africa. *New York Times*. http://www.nytimes.com/2008/01/31/world/africa/31safrica.html. Accessed 13 April 2011.

Dammert, A., Gallardo, J., & García, R. (2005). *Reformas Estructurales en el Sector Eléctrico Peruano* (Working Document No.5). Lima: OSINERG. www.osinerg.gob.pe/newweb/uploads/Estudios_Economicos/DT05-OEE-OSINERGa1.pdf. Accessed 13 April 2011.

Gómez, M. F., & Silveira, S. (2010). Rural electrification of the Brazilian Amazon: Achievements and lessons. *Energy Policy, 38*(10), 6251–6260.

Gratwick, K. N., & Eberhard, A. (2008). An analysis of independent power projects in Africa: Understanding development and investment outcomes. *Development Policy Review, 26*(3), 309–338.

IEA (International Energy Agency). (2008). *World energy outlook 2008*. Paris: OECD/IEA.

Kanagawa, M., & Nakata, T. (2008). Assessment of access to electricity and the socio-economic impacts in rural areas of developing countries. *Energy Policy, 36*(6), 2016–2029.

Kooijman-van Dijk, A. L., & Clancy, J. (2010). Impacts of electricity access to rural enterprises in Bolivia, Tanzania and Vietnam. *Energy for Sustainable Development, 14*(1), 14–21.

Wilson, M., Besant-Jones, J., & Audinet, P. (2010). *A new slant on slopes: Measuring the benefits of increased electricity access in developing countries*. Washington, DC: World Bank/Energy Sector Management Assistance Program (ESMAP).

World Bank. (1993). *The world bank's role in the electric power sector: Policies for effective institutional, regulatory, and financial reform* (Policy Paper). Washington, DC: World Bank.

World Bank. (1999). *Reforming Bolivia's power sector* (Précis No. 192). Washington, DC: World Bank Operations Evaluation Department.

Financing Renewable Energy*

Sujata Gupta

Abstract Renewable energy sources are inexhaustible, have minimal adverse impacts and could play a pivotal role in meeting the world's future energy demands in a sustainable manner. This chapter discusses opportunities and strategies to exploit the potential of renewables, and the market challenges that affect investments. Financing is often difficult for renewable energy projects due to the high risk perception associated with new technologies and the resulting high costs of capital. To overcome these barriers, and to promote the deployment of renewable technologies, the Asian Development Bank has adopted a number of innovative financing mechanisms, including clean energy funds, credit lines and first-loss guarantees.

1 Introduction

Renewable energy resources are inexhaustible, have minimal adverse impacts and could play a pivotal role in meeting the world's future energy demands in a sustainable manner. This chapter discusses the potential of renewables and the market challenges that impact investments. It then examines the opportunities and strategies to exploit their potential, related issues that need to be addressed, and the support of the Asian Development Bank (ADB) to renewable energy projects.

Despite the large potential of renewable forms of energy that are suitable for a variety of end-uses, there are significant barriers that prevent their exploitation. Innovative mechanisms will be required to overcome these barriers. For example, at

*Disclaimer: the views and opinions presented here do not necessarily reflect the views and opinions of the Asian Development Bank.

S. Gupta (✉)
Asian Development Bank, 6 ADB Avenue, Mandaluyong, Metro Manila 1550, Philippines
e-mail: sgupta@adb.org

F.L. Toth (ed.), *Energy for Development: Resources, Technologies, Environment*, Environment & Policy 54, DOI 10.1007/978-94-007-4162-1_14,

Box 1 Terms in Power Sector Financing

Long-term preferential feed-in tariff is the cost plus based price per unit of renewable electricity from private generators. It is typically paid to renewable energy developers so they can compete on price with conventional (and less costly) forms of power.

Guaranteed power off-take commitment is an agreement to purchase all or a substantial part of the electricity produced by a project, which would provide an assured revenue stream to the project.

Subordinated debt ranks after other debts should a company fall into insolvency or be closed. Such debt is referred to as subordinate, because the debt providers (the lenders) have subordinate status in relationship to the normal debt.

Senior debt takes priority over other unsecured or otherwise more "junior" debt owed by the issuer. Senior debt has higher priority in the issuer's capital structure than subordinated debt. In the event the issuer goes bankrupt, senior debt theoretically must be repaid before other creditors receive any payment.

Limited recourse debt is a debt in which the lender's claims are limited in the event of default.

Tenor is the remaining duration for the repayment of a loan or contract or the initial term length of a loan.

First loss guarantee pays for the initial loss incurred up to the maximum guaranteed amount. The lender will only suffer losses when the total loss exceeds the guaranteed amount.

present it is difficult for many projects to obtain financing due to the perceived high level of risk associated with new technologies and the resulting high costs of capital investments. Therefore, the initial development stages of such options require effective implementation of supportive policies and regulations that provide incentives such as long-term preferential feed-in tariffs and guaranteed power off-take commitments (Box 1).

2 Renewable Energy: Potentials, Opportunities and Challenges

Renewable energy resources, such as wind, solar and hydropower, are well distributed around the world. Figure 1 shows the global irradiation levels relevant for solar energy. Table 1 presents the potential of different forms of renewables, in

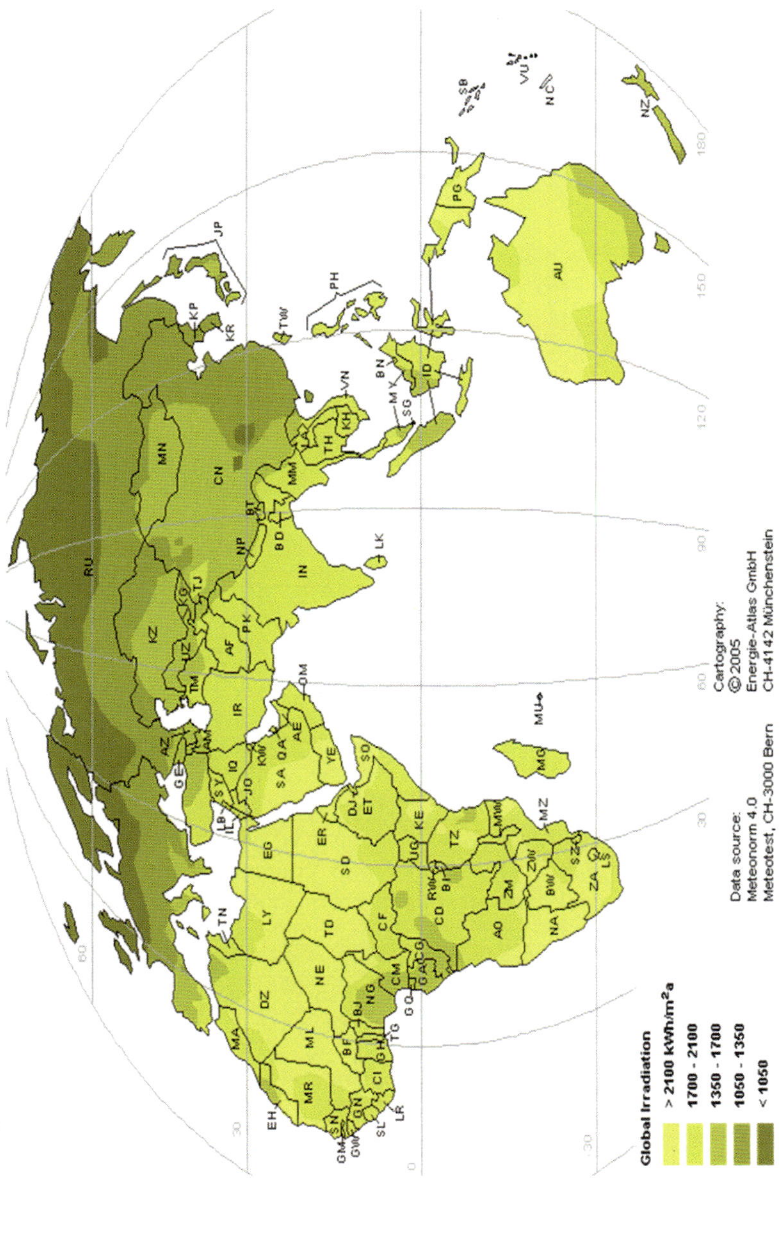

Fig. 1 Dispersed sources: Solar energy potentials are high in many parts of the developing world (Source: Energie-Atlas GmbH 2005)

Table 1 Renewable energy: physical, technical and socio-economic potentials (exajoules/year; 1 EJ = 10^{18} J) (Sources: UNDP 2000, 2004)

	Physical (theoretical)	Technical	Socio-economic	Currently exploited
Hydropower	147	50	28	10
Biomass	2900	276–446	?	50
Solar	3.9 million	1,600–50,000	?	0.2
Wind	6,000	640	231	0.2
Geothermal	140 million	5,000	500	2.1
Marine energy	7,400	10	?	0.002
Total	**>143 million**	**>7,500**	**?**	**62.4**

terms of the amounts of energy theoretically available, technically exploitable, and socially and economically feasible (see also Rogner 2012). The last column gives the energy currently obtained from renewable sources; clearly, despite the significant potential, very little has yet been harnessed. In some cases, such as marine energy, the barriers are technical, while in others, such as solar power, they are social and economic. So far, only a fraction of the technically and socio-economically feasible potential of these resources has been realised.

Wind energy technology, for instance, is now well established, and has been commercially viable for the last 5–10 years. Yet only 0.2 exajoule (EJ = 10^{18} J) of the potential 231 EJ/year is being harnessed. One of the main barriers preventing the further diffusion of this technology is financing.

Renewable sources of energy have several advantages. They contribute to a country's energy security and energy price stability by reducing its dependence on energy imports. Renewable energy installations are not dependent on fuel imports, and their operations are not affected by supply disruptions. Countries that have a high percentage of renewables in their energy portfolio are less exposed to fluctuations in global energy prices. Indeed, over time, a greater reliance on renewable energy supplies would reduce the dependence on conventional energy and international markets, leading to more stable world energy prices. Renewable resources are decentralised, providing the best options for improved access to clean energy for people in remote areas, especially in developing countries. Further, it has been established that on a life cycle basis, renewable energy options emit fewer pollutants than conventional fuels, with benefits at the household, local and global levels.

Despite these benefits, renewable options face many challenges that make it difficult for them to compete with conventional sources of energy. These include limited and intermittent supplies, the high up-front capital investment required, problems in operation and maintenance (O&M) due to their dispersed and small scale, and difficulties in obtaining financing.

Renewable energy *supplies* are highly dependent on nature, particularly the weather, and supply–demand mismatches can occur often. Maintaining reliable power supplies therefore requires additional backup or storage systems, all of which add to the cost. Further, to produce the same amount of power, installations such as solar parks or wind farms require much larger areas of land than conventional power plants. At current prices and output (or efficiency) levels, renewable technologies are expensive, although prices have fallen sharply as global capacity has increased. Renewable technologies are still in the early stages of market penetration, and their prices relative to fossil fuel-based technologies are likely to decline even further in the future.

Another major challenge is *scale*. As it becomes possible to exploit economies of scale, costs will decline significantly, as has been demonstrated in the case of wind turbines and, more recently, solar power. According to McKinsey (2008), for each doubling of cumulative capacity installed, costs have decreased significantly, by 23% for solar photovoltaics (over the period 1975–2003), 13% for wind (1981–2001), 15% for ethanol (1976–1996) and 3% for solar thermal (1985–1991).

Renewable energy technologies are capital intensive, and the problems of low capacity utilisation and the high cost per megawatt (MW) are further compounded by the fact that the *capital investment* is front-loaded, i.e. most of the costs are incurred at the start of a project. Given that these technologies are evolving and in their current form have relatively short operating track records, and performance is highly site specific, most projects are perceived to be financially risky. Although fuel and O&M costs are very low, *financing* these installations requires long-term loans that can be repaid through the savings in fuel costs. Therefore, promoting renewables requires considerable resources with very long payback periods for projects with no proven track record. Banks have a high risk perception of such projects and are hesitant to finance renewable energy projects.

Most renewable technology options are small scale and hence the transaction costs are significant. For instance, to develop a renewable capacity equivalent to that of a single 300 MW coal-fired power plant would require at least 10–50 projects, which in turn would involve several providers; and power purchase agreements (PPAs) would have to be signed for each facility. All this adds to the transaction costs associated with renewable options, making them less commercially attractive and less bankable than conventional technologies.

The constraints on the development of renewables are further compounded by subsidies enjoyed by conventional energy projects. According to the *World Energy Outlook* (IEA 2010), the implicit and explicit subsidies for fossil fuels amounted to US$312 billion in 2009. In addition, by not including the additional costs associated with conventional energy, such as adverse health impacts, or by not internalising the benefits or positive externalities that come from the use of renewables, the market is distorted, further undermining their competitiveness.

Investments in renewables face several market challenges. The financial sector needs to adapt, and to design suitable instruments as well as innovative mechanisms that will address the technological and financial barriers, and unlock the large potential of renewable energy projects. It is also important that the financial institutions develop their own capacity, at both management and operational levels, so that they are better equipped to understand, analyse and support renewable technologies. It is equally important to bring down the costs of new technologies, perhaps by exploiting economies of scale, although the high costs are preventing these options reaching scale. This has led to a vicious circle that is inhibiting the large-scale deployment of renewables.

Table 2 Examples of renewable energy targets in Asian developing countries (Source: REN21 2010)

Country	Targets
China	Renewable capacity: 362 GW by 2020, including 300 GW hydro, 30 GW wind, 30 GW biomass, and 1.8 GW solar PV/CSP, although increased targets to 150 GW wind and 20 GW solar PV/CSP by 2020 exist as draft or unofficial targets; solar hot water: 150 million m^2 by 2010 and 300 m^2 by 2020
India	Renewable capacity: 12.5 GW added 2007–2012; 15% share of added power capacity 2002–2022
	Solar PV and CSP: 1.1 GW by 2013, 10 GW by 2017, 20 GW by 2022
	Wind power: 9 GW added 2007–2012
	Small hydro: 1.4 GW added 2007–2012
	Biomass/cogeneration: 1.7 GW added 2007–2012
	Waste-to-energy: 0.4 GW added 2007–2012
	Solar hot water: 15 million m^2 by 2017; 20 million m^2 by 2022
	Rural lighting systems: 20 million by 2022
Indonesia	Geothermal: 6 GW; biomass: 810 MW; wind power: 255 MW; solar PV: 80 MW (all by 2025)
Pakistan	Renewable capacity: 5% by 2030
Philippines	Renewable capacity: 4.5 GW added during 2003–2013
Thailand	Solar PV: 0.055 GW by 2011, 0.095 GW by 2016, 0.500 GW by 2022
	Wind: 0.115 GW by 2011, 0.375 GW by 2016, 0.800 GW by 2022
	Hydro: 0.185 GW by 2011, 0.281 GW by 2016, 0.324 GW by 2022
	Biomass: 2.8 GW by 2011, 3.22 GW by 2016, 3.7 GW by 2022
	Biogas: 0.06 GW by 2011, 0.09 GW by 2016, 0.12 GW by 2022
Vietnam	3% of total primary energy by 2010, 5% by 2020 and 11% by 2050

PV photovoltaics, *CSP* conventional solar power

Developing countries have undertaken impressive initiatives to promote renewables (see Table 2). But what they can achieve on their own is limited, given that the renewable technologies are not yet established, they are relatively expensive and capital intensive, and they compete with other social priorities for the limited capital available.

In Asia, the energy sectors in China and India have grown rapidly in recent years. In 2008 alone, China added the equivalent of three conventional 500–600 MW power plants each week. India lags far behind China, but has added two 500 MW power plants each month in 2007. Both countries have set impressive targets for renewable capacity, but these targets are much lower than those for the conventional energy sector and given lower plant load factors for renewables, their contribution in the generation mix is much smaller. The above statistics show that a window of opportunity to build significant renewable capacity exists now, while developing countries are adding to their power generation capacities. Once created, these capacities will operate for the next 30–40 years. External, non-market stimuli are required to exploit this opportunity to accelerate the penetration of renewable energy.

Developing countries are well-endowed with renewable resources, yet projections by the International Energy Agency indicate that coal will remain the world's primary fuel even in the year 2030 (IEA 2010). Between 2008 and 2035, global primary energy demand will increase by 36%, and electricity demand by 2.2% per year, more than 80% of which will be in non-OECD countries. The IEA (2010) also projects that in 2035, 55% of the total electricity generated will be produced from fossil fuels, mainly coal and natural gas, thus firmly establishing their dominance in the energy mix. This implies that the transition to renewables will be rather slow, and accelerating their penetration will require an external impetus in the form of international financial and technical support to expand the renewable energy capacity in developing countries. Such support will have universal benefits, including reductions in the cost of technologies due to economies of scale, new business development and environmental benefits such as lower greenhouse gas emissions.

3 Investments in Renewable Energy

The potential for investment in renewable energy is very large in developing countries. This includes not only the installation of generating capacity but also the creation of manufacturing capacity, which in turn can lower the costs of renewable options because of the relatively low labour costs and transport requirements. Greater deployment of renewable technologies will lead to job creation in developing countries (see Table 3) and to lower technology costs, thus opening up new avenues for growth and profit for the technology worldwide.

Accelerating the share of renewables in the energy mix is a win-win proposition, but a coherent global strategy is needed to exploit their potential, one that looks beyond national policies. Although some small and significant steps have been taken, they have so far been inadequate. The efforts to promote renewable energy have to be global, and will require additional capital and support since these are high-cost options and developing countries are capital constrained. Some highly sophisticated technologies are emerging, and most of them are in the developed world (see Gibbons 2012).

Developing countries need to build new generation capacity, and with the right kind of technical and financial support a substantial part of this can be based on renewables. Supporting and developing renewable energy options is one of the best ways to meet the global greenhouse gas emission targets. It is necessary to regard support to renewable energy projects not purely as grants, but more as facilitating or leveraging mechanisms that can assist 'first movers', transform pilots into commercial-scale projects, and reduce the perceived level of risk associated with new technologies. External support for renewables is required because of market imperfections (such as the incomplete internalisation of the costs associated with fossil energy), the limited support infrastructures for renewables, and the need to cover the potential risks of first movers. This support should be designed to address these issues and other associated risks in order to accelerate the penetration of renewable technologies.

So far, renewables have been considered at a small scale, which has not been very useful as it has led to limited penetration of the various options. It is necessary to scale up renewable options that will break the vicious cost–scale loop. For example,

Table 3 Estimates of jobs in renewable energy industries in 2009 (Source: REN21 2010)

Industry	Estimated jobs worldwide	Selected national estimates
Wind power	>500,000	Germany 100,000; United States 85,000; Spain 42,000; Denmark 22,000; India 10,000
Solar hot water	~300,000	China 250,000
Solar PV	~300,000	Germany 70,000; Spain 26,000; United States 7,000
Solar thermal	~2,000	Spain 1,000; United States 1,000
Biomass	–	Germany 110,000; United States 66,000; Spain 5,000
Hydropower	–	Europe 20,000; United States 8,000; Spain 7,000
Geothermal	–	Germany 9,000; United States 9,000
Biofuels	>1.5 million	Brazil 730,000 (in sugar cane and ethanol production)
Total	>3 million	

if China needs to build three power plants each week, it is not sufficient to provide solar lanterns to a village. According to the IEA's projections, under a business as usual scenario, coal will remain the dominant fuel. Therefore, ways to scale up investments have to be devised such that renewable energy projects can increase their share in the generation mix and make the transition from kilowatt and a few megawatts scale to a gigawatt scale. This requires exploiting the technical and economic potential more effectively. It is essential to create a supply side that can deliver technology options that are efficient, offer reliable performance levels, and are reasonably priced. Expansion of scale requires adequate manufacturing facilities that can guarantee quality products, or else national or regional policies are likely to suffer setbacks, as was the case in Germany when the government decided to promote solar power, and the price of photovoltaic (PV) panels increased. This market distortion was the result of insufficient global PV production capacity.

In its annual global status report, the Renewable Energy Policy Network for the twenty-first century (REN21) notes that the growth in investments has slowed despite the 'green stimulus' efforts by many major economies (REN21 2010). It reports a 6% drop in global investments in utility-scale additions, from US$108 billion in 2008 to US$101 billion in 2009 (see Fig. 2).

Recognising the issues that need to be addressed to accelerate the development of renewable energy technologies, immediate interventions are required at the international level; waiting for the market to play out is not an option. Both developed and developing countries, the private sector, the financial sector and multi- and bilateral development agencies all have clear roles to play.

4 ADB Support to Renewable Energy

The Asian Development Bank has responded to the need to promote renewables through its Clean Energy and Environment Program (see Fig. 3). This chapter discusses a number of initiatives launched as part of this comprehensive strategic effort,

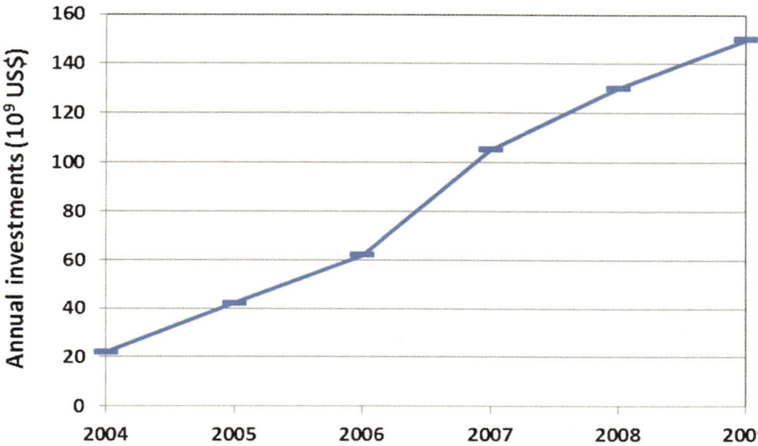

Fig. 2 Global investments in new renewable energy capacity, 2004–2009 (Source: REN21 2010)

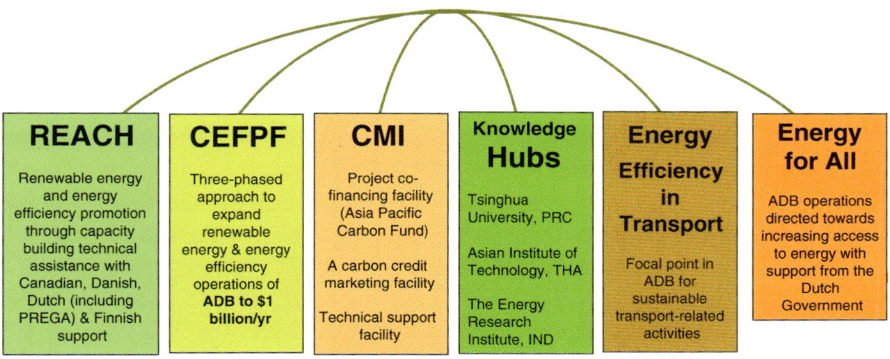

Fig. 3 ADB's Clean Energy and Environment Program (Source: ADB 2011a; Note: *CEFPF* Clean Energy Financing Partnership Facility, *CMI* Carbon Market Initiative, *PREGA* Promotion of Renewable Energy, Energy Efficiency and Greenhouse Gas Abatement programme; *PRC* People's Republic of China, *THA* Thailand, *IND* India)

and information on the others is available from the ADB website (ADB 2011a). The ADB has established a Clean Energy Financing Partnership Facility (CEFPF), the main objective of which is to build the capacity of developing member countries to implement national strategies for exploiting renewables and improving energy efficiency in their economies. In 2008, the ADB set itself the target of investing US$1 billion per year to support clean energy initiatives.

Figure 4 illustrates other objectives of the CEFPF, such as minimising the credit and technological risks associated with renewable energy projects. Most important, the ADB has recognised that in order to bring about change, it needs to develop its own internal capacity, and has taken steps to enhance the knowledge of its operational staff involved in scaling up clean energy initiatives and in monitoring and evaluation.

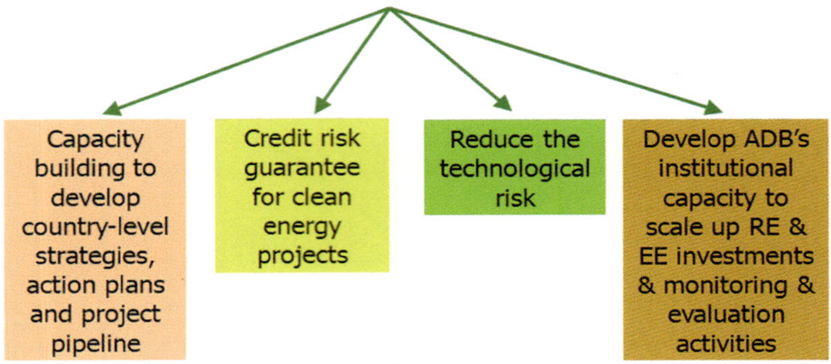

Fig. 4 The ADB's Clean Energy Financing Partnership Facility (CEFPF) (Source: ADB 2011a; Note: *RE* renewable energy, *EE* energy efficiency)

Table 4 Overview of ADB's clean energy investments, 2003–2010 (Source: ADB 2011a)

Year	Clean energy investments (US$ million)
2003	226
2004	306
2005	757
2006	657
2007	668
2008	1,743
2009	1,313
2010	1,756

Based on assessments of its developing member countries, the Bank then developed a project pipeline based on the absorptive capacity in each case. The ADB will now collaborate with other institutions and donors/bilateral agencies in developing the capacities of these member countries, and design instruments to reduce the financial and technological risks associated with renewable technologies.

The ADB has consistently achieved its target of investing at least US$1 billion annually in clean energy (Table 4). Of the total lending portfolio of US$11 billion approved in 2008, US$3.1 billion was devoted to energy projects, of which US$1.7 billion (55%) to clean energy (see Fig. 5). Starting in 2013, the target will be increased to US$2 billion per year, in a continued bid to accelerate low-carbon growth and reduce greenhouse gas emissions.

5 Financing Instruments Adopted by the ADB

Another ADB intervention to promote clean energy projects is the Carbon Market Initiative (see Fig. 6), which comprises three elements:

- A Technical Support Facility, which helps to identify and develop the carbon emission reduction components of projects;

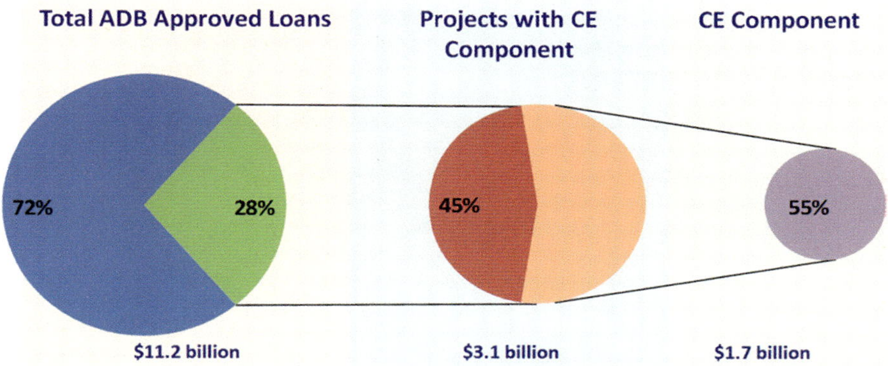

Fig. 5 The ADB's clean energy (CE) investments in 2008 (Source: ADB 2011a)

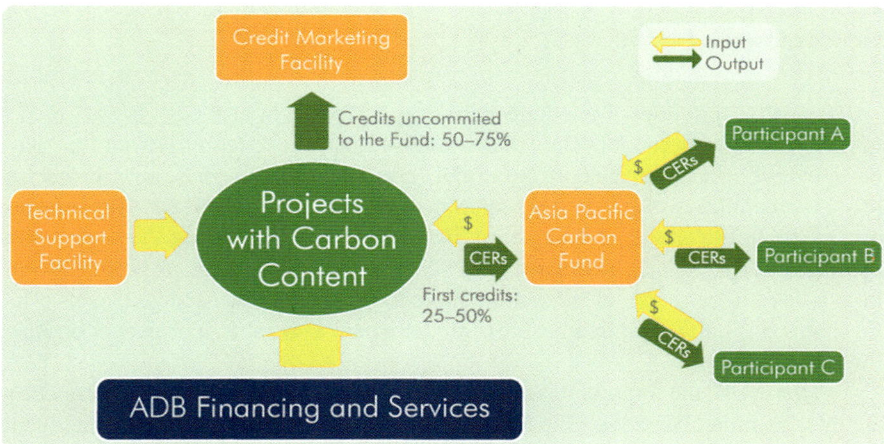

Fig. 6 The three components of the ADB's Carbon Market Initiative (Source: ADB 2010; Note: *CERs* certified emission reductions)

- A Credit Marketing Facility, which helps to market part of the certified emissions reductions (CERs) credits generated by clean energy projects; and
- Most important in terms of project financing, the Asia Pacific Carbon Fund (APCF), which buys upfront a percentage (25–50%) of the CERs generated by projects.

Typically, any project, including those that access carbon financing, goes through two distinct phases – construction and commercial operation. The CERs or carbon credits generated during the commercial operation phase are sold to an entity either through an existing purchase agreement or on the spot market. Thus, the purchasing entity is in effect 'financing' the project when the carbon credits are delivered to it. This conventional business model is called the 'pay-on-delivery' model (see Fig. 7).

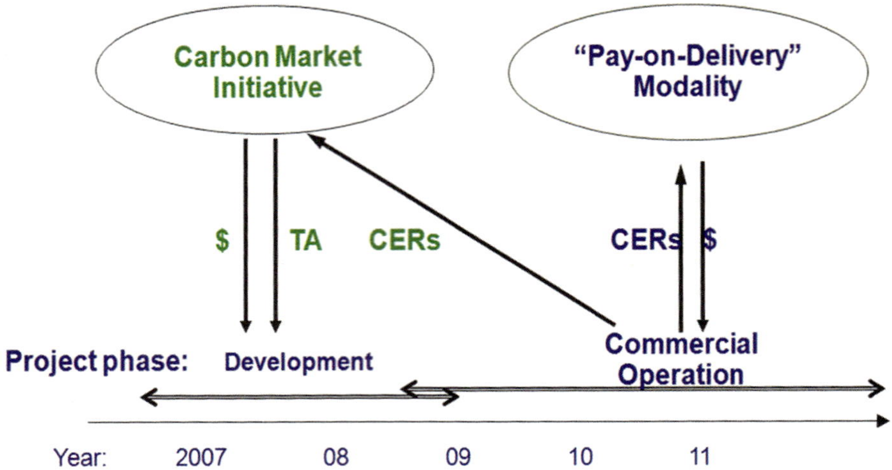

Fig. 7 The ADB's risk sharing model (Source: ADB 2010; Note: *TA* technical assistance, *CERs* certified emission reductions)

The APCF provides immediate financing for projects and has also adopted a unique risk-sharing model. The ADB initiative operates differently from the "pay-on-delivery" model because in the initial (pre-construction) phase of a project, it provides technical assistance to develop the project and buys and pays up-front for part of the future credits. This up-front investment is repaid during the commercial operation of the project. In effect, the APCF shares the project risk prior to construction, a risk it faces in the event that the project operates below expected levels. This mechanism is equivalent to a business model where a 'loan' to a project is serviced through CERs.

For the ADB, the important issue is to identify the right technologies and accelerate their commercialisation. The latter requires suitable financing models that blend public and private resources. The main purpose of public resources is to create an enabling environment that alleviates the barriers that are preventing private resources from flowing to the identified technologies. Given the magnitude of the market and the funding gap for promoting clean energy technologies, a substantial increase in concessional support from public sources, channelled through multilateral and bilateral financial institutions, is required to mitigate the technological risk and to support technology transfer.

In addition to the APCF, other financial instruments could assist in engaging the private sector in clean energy projects. One option is to increase the share of equity for clean energy technologies to attract commercial debt funding for the balance. Greater proportions of equity help in building the confidence of commercial financiers in these technologies. As there is a scarcity of equity capital, the *incremental equity* has to come from sources that are willing to take the higher level of risk associated with renewable technologies. An alternative to higher level of equity participation is *subordinated debt,* which is a loan that will be paid only after other senior debt has

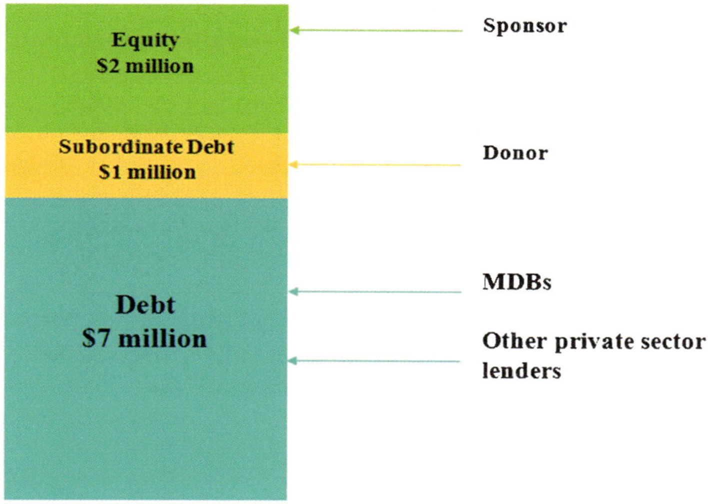

Fig. 8 Subordinated debt (Note: *MDBs* multilateral development banks)

been repaid. This again encourages commercial banks to provide senior debt financing to clean energy projects as their loans are sufficiently collateralized.

Figure 8 shows an example of subordinated debt where the sponsor provides equity of US$2 million, the donor funds contribute subordinated debt of US$1 million, and the rest of the financing is obtained on commercial terms. In the event that a project is unable to service its debt, the commercial financiers are assured that they would be repaid before the donor's subordinated debt and before any returns to the equity holders. The equity holders absorb the maximum risk and are the last to be repaid.

Other options include *credit lines* and *loans with in-built incentives* such as performance bonuses or interest rate step-downs if certain milestones or targets are achieved. For example, a credit line is extended to a commercial bank to on-lend to renewable energy projects on the condition that once a minimum specified lending to such projects has been achieved, the initial rate of interest at which money was lent to the bank would be lowered. Therefore, there is an incentive for the bank, at the grassroots level, to increase lending for renewable energy projects. At the same time, there is a built-in check in this mechanism, as the on-lending bank always has a stake in the transaction because ultimately it is investing its own money in the project. The bank has to ensure that a thorough assessment ('due diligence') is done before lending to any project. However, reducing the rate of interest implies that there is some element of subsidy, the cost of which has to be borne by 'concessional' sources.

Another instrument is a *partial risk guarantee*, where the potential inability of a project to repay due to certain unique characteristics is covered, fully or partially, by concessional resources. This mechanism enhances the credit quality of the project

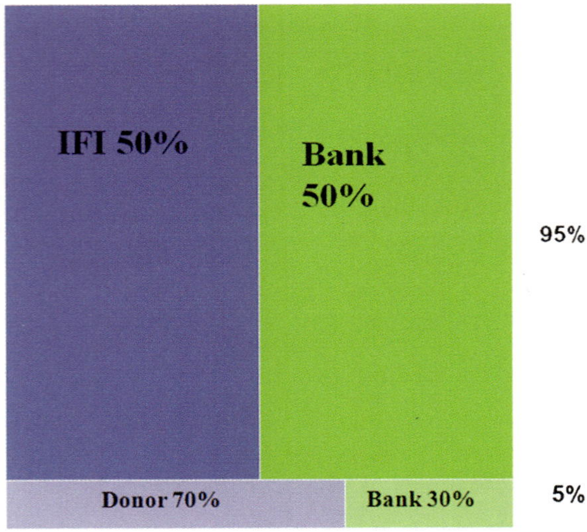

Fig. 9 Risk sharing in a bank's portfolio of renewable energy projects (Note: *IFI* international financial institution)

and enables financing, possibly at a lower rate than it would have obtained in the market given the perceived risk attached to it.

The ADB recently in 2011 approved a *partial credit guarantee* (PCG) for solar projects in India (ADB 2011b). Some solar projects have technical and resource related risks that cannot be easily mitigated for the first few projects, so banks remain wary of lending to them and/or of lending beyond their existing corporate relationships. The Bank has therefore created a facility to issue PCGs amounting to up to US$150 million of principal, to promote foreign and local commercial bank lending to solar power generation projects in India. The facility will support multiple private sector solar projects up to a maximum size of 25 MW. Under this facility, the ADB will issue PCGs to guarantee scheduled payments of principal and interest under loans to be provided by foreign or local commercial banks. The facility will share the credit risks with commercial lenders. The facility has the twin objectives of (i) making limited recourse debt financing available on reasonable terms and conditions, and (ii) extending the tenor of loans to solar projects. The ADB will issue PCGs for solar projects in India to international and local lenders to cover up to 50% of the unpaid loan and interest amount. The partial cover by ADB of all the risks associated with the solar power projects would substantially improve the risk rating of the projects making them more bankable. The ADB is in discussion with donors to obtain concessional resources to reduce the guarantee fees.

Figure 9 shows an example of risk sharing in which an international financial institution (IFI) gives a bank guarantee blended with donor funds to cover part of the first loss. In the example, the first 5% loss on the project is covered up to 70% by donor or concessional funds, and the commercial bank's exposure is only 30%. Beyond this first 5%, the commercial bank bears 50% of the loss and the IFI the other 50%. The probability of loss beyond the first initial losses is much lower, so if a donor shares a higher percentage (could be 100% also) of the first loss, it significantly improves the bankability of the project. The commercial bank retains a stake in the project at all levels but with an overall lower level of risk exposure.

However, all of these risk reduction and risk-sharing mechanisms – incremental equity, bonus or interest rate step-downs and credit guarantees – need to be financed by donor or concessional funds, and this is where the international community plays an important role. Using these mechanisms, the penetration of clean technologies is accelerated while, at the same time, through the successful demonstration of the projects, the excessive perceived risk is eliminated.

6 Conclusions

Energy is a prerequisite for poverty alleviation, as pointed out in other chapters in this volume (e.g. Yumkella 2012; Brew-Hammond 2012). The demand for energy is best met from renewable sources that are sustainable, improve access to energy for people in remote areas and enhance energy security. However, for renewable energy to make any significant contribution, or to be a serious alternative to conventional energy, a massive scale-up of interventions, investments and businesses is required. The Asian Development Bank has adopted various innovative financing mechanisms to support the dissemination of renewable energy technologies, including clean energy funds, credit lines and partial credit guarantees.

Acknowledgment The author would like to gratefully acknowledge the research support provided by Disha Agarwal, MTech student, TERI University, New Delhi, India.

References

ADB (Asian Development Bank). (2010). *Carbon market program*. http://www.adb.org/Documents/brochures/cmp/brochure-cmp.pdf

ADB (Asian Development Bank). (2011a). *Clean energy: ADB's clean energy program*. www.adb.org/Clean-Energy/default.asp

ADB (Asian Development Bank). (2011b). *India solar generation guarantee facility*. www.adb.org/About/Private-Sector/India-Solar-Guarantee-Facility.asp

Brew-Hammond, A. (2012). Energy: The missing Millennium Development Goal. In F. L. Toth (Ed.), *Energy for development: Resources, technologies, environment* (pp. 35–43). Dordrecht: Springer.

Energie-Atlas GmbH. (2005). *Global irradiation map*. Munchenstein: Energy-Atlas GmbH.

Gibbons, J. H. (2012). Technology and innovation. In F. L. Toth (Ed.), *Energy for development: Resources, technologies, environment* (pp. 141–148). Dordrecht: Springer.

IEA (International Energy Agency). (2010). *World energy outlook 2010*. Paris: IEA.

McKinsey. (2008). *The carbon productivity challenge: Curbing climate change and sustaining economic growth*. Washington, DC: McKinsey Global Institute. www.mckinsey.com/mgi/reports/pdfs/Carbon_Productivity/MGI_carbon_productivity_full_report.pdf

REN21 (Renewable Energy Policy Network for the 21st Century). (2010). *Renewables 2010 global status report.* Paris: REN21 Secretariat.
Rogner, H.-H. (2012). Energy resources. In F. L. Toth (Ed.), *Energy for development: Resources, technologies, environment* (pp. 149–160). Dordrecht: Springer.
UNDP (United Nations Development Programme). (2000). *World energy assessment: Energy and the challenge of sustainability.* New York: UNDP/UN Department of Economic and Social Affairs/The World Energy Council. www.undp.org/energy/activities/wea/drafts-frame.html
UNDP (United Nations Development Programme). (2004). *World energy assessment: Overview 2004 update.* New York: UNDP/UN Department of Economic and Social Affairs/The World Energy Council. www.undp.org/energy/docs/WEAOU_full.pdf
Yumkella, K. K. (2012). Multilateralism and energy for development. In F. L. Toth (Ed.), *Energy for development: Resources, technologies, environment* (pp. 45–56). Dordrecht: Springer.

Fossil Fuels and Carbon Capture and Storage

Keith Burnard and Sean McCoy

Abstract Reducing CO_2 emissions, including those from the energy sector, presents a major challenge to the world at large. Fossil fuels provide two-thirds of the world's electricity, with coal, in particular, the fuel of choice for new electricity generation plant in rapidly emerging economies such as China and India. As a result, carbon capture and storage (CCS) is an essential component of any portfolio of measures to reduce CO_2 emissions at least cost and to prevent the damaging consequences of climate change. The successful deployment of CCS will depend on progress in many areas – policy, legal, financial, technical, market and organisational – and, not least, on public acceptance of the technology.

1 Introduction

Reducing the threat of climate change resulting from increasing carbon dioxide (CO_2) concentrations in the atmosphere presents a major challenge to the energy community and to the world at large. The G8 leaders, following their 2010 Summit in Muskoka, Canada, reaffirmed that '*... As we agreed in L'Aquila, we recognize the scientific view that the increase in global temperature should not exceed 2 degrees Celsius compared to pre-industrial levels*', and further stated that '*Achieving this goal requires deep cuts in global emissions...*' (G8 2010).

To keep the temperature rise below 2°C, it is anticipated that global CO_2 emissions will need to be reduced by between 50% and 80% compared to 2000 levels by 2050. In a recent publication, 'Energy Technology Perspectives 2010: Scenarios and

K. Burnard (✉) • S. McCoy
International Energy Agency, 9 rue de la Fédération, 75739 Paris Cedex 15, France
e-mail: keith.burnard@iea.org; sean.mccoy@iea.org

F.L. Toth (ed.), *Energy for Development: Resources, Technologies, Environment*, Environment & Policy 54, DOI 10.1007/978-94-007-4162-1_15,

Strategies to 2050', the International Energy Agency (IEA) focused on two main scenarios (IEA 2010a). The Baseline scenario offers a view of developments to 2050 that might occur if no new energy and climate policies were introduced, while the BLUE Map scenario examines the least-cost means to halve global CO_2 emissions by 2050. The IEA analysis suggests that a portfolio of technologies and policy measures are required to reduce global emissions and help avoid the most serious impacts of climate change. The technology portfolio includes measures to improve the efficiency with which we use energy, a switch to lower carbon fuels, to increase the energy supplied from renewable sources and nuclear, and the widespread application of carbon capture and storage (CCS). Analysis has indicated that CCS contributes significantly to the least-cost route of reducing emissions and stabilising CO_2 concentrations in the atmosphere (IEA 2010a).

This chapter examines the role of fossil fuels in power generation and the means by which emissions can be reduced. In many countries, a substantial proportion of the electricity is generated from fossil fuels, so regional contributions to energy-related CO_2 emissions will be discussed. Strategies and technologies need to be developed to limit the CO_2 produced and, where it is produced, to reduce the amount emitted to the atmosphere. Although CCS is recognised as an important component of a viable CO_2 mitigation strategy, the technology is still in its infancy. Capture technologies have yet to be demonstrated at commercial scale on a power generation plant, and the accompanying CO_2 transport infrastructure and storage sites remain to be developed. In this chapter the potential of CCS technologies is assessed, the challenges they face explored and the promised benefits examined.

2 Cleaner Fossil Fuel Technologies

For deep cuts to be made in global CO_2 emissions, it is essential to address power generation. Electricity production accounts for 35% of total global fossil fuel use and around 41% of total energy-related CO_2 emissions. Almost 70% of electricity is generated from fossil fuels and this share has increased steadily since 1990, as shown in Fig. 1. Coal is the most important energy source, producing 41% of the world's electricity in 2008. Whereas the contribution from gas lay at 21% in 2008, having grown rapidly over the previous 20 years, oil's share was just 5% (IEA 2010b).

At present, the use of coal as a fuel source for power generation shows no signs of abating. In the rapidly expanding economies of China and India, for example, coal is fuelling the overwhelming majority of their new power generation capacity. With increased coal usage, however, come a number of challenges. Apart from CO_2 emissions, emissions of SO_2, NO_x and particulates – the so-called traditional pollutants – from coal combustion must also be addressed; they can lead to significant health and environmental effects. These were addressed extensively throughout the 1960s, 1970s and 1980s to the present day, and technologies have been developed that can reduce the concentrations of these pollutants to very low levels in power plant flue gas. In more recent years, effort has been focused on reducing the release

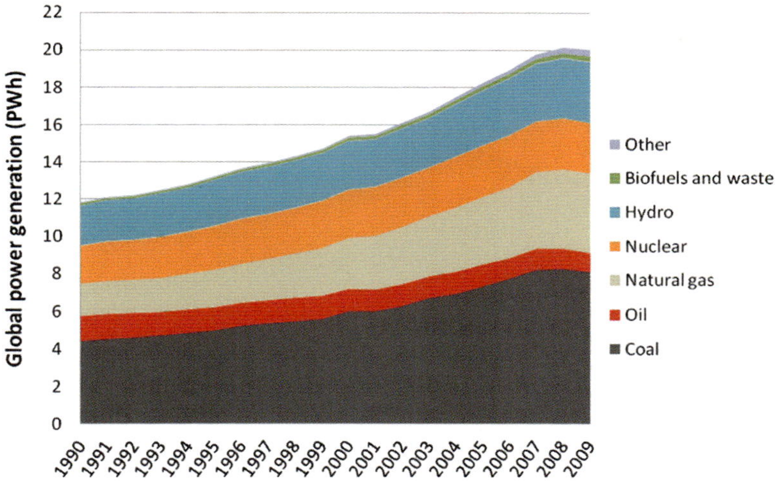

Fig. 1 Historical role of fossil fuels in power generation, 1990–2009

of heavy metals into the environment from coal combustion, and this effort is continuing. Legislation and its effective enforcement are fundamental to ensure that technologies developed to reduce pollutant emissions are successfully deployed.

Following recognition of the role of anthropogenic emissions on global warming, however, emissions of greenhouse gases are now the subject of global concern. As well as being the dominant fuel source for power generation, coal is also the most carbon intensive. For this reason, it receives most attention in the development of greenhouse gas mitigation strategies. CO_2 emissions have risen steadily in recent years, reaching 29 gigatonnes (Gt) in 2008, 41% of which were from power generation (IEA 2010b). In the absence of new CO_2 mitigation policies, the IEA estimates that emissions could rise to 57 Gt by 2050, 40% of them from power generation (IEA 2010a).

The United Nations Intergovernmental Panel on Climate Change (IPCC) has concluded that increases in emissions of this magnitude would result in dramatic climate change, including an increase in global average surface temperature of more than 6°C (IPCC 2007). To avoid such dangerous climate change, emissions of CO_2 to the atmosphere must be reduced. Limiting temperature rise to 2–3°C – and hopefully avoiding the worst impacts of climate change – will require reducing substantially the CO_2 concentration in the atmosphere. This premise underpins the IEA's BLUE Map scenario. In this scenario, emissions in 2050 are reduced to 14 Gt, around half the level emitted in 2005 and 43 Gt lower than projected in the Baseline scenario. Achieving these CO_2 emission reductions will require the development and deployment of a portfolio of measures to improve energy efficiency and to promote low-carbon technologies, as illustrated in Fig. 2.

In the BLUE Map scenario, contributions to the reduction in CO_2 emissions will come from CCS, renewables, nuclear and energy efficiency improvements. Around one-third of the 43 Gt CO_2 reductions needed in 2050 would come from power

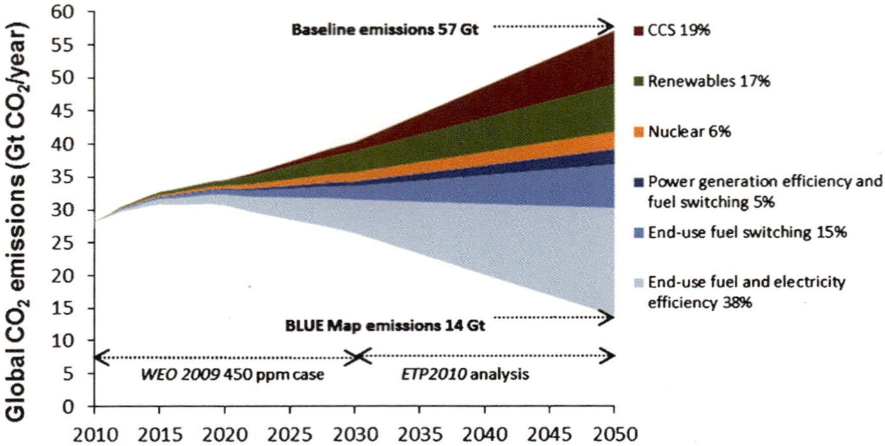

Fig. 2 Key options for reducing CO_2 emissions under the BLUE Map scenario (Source: IEA 2010a)

generation, which is largely decarbonised, and the remainder from the construction, transport and industry sectors. For all of these sectors, the challenges cannot be underestimated; unprecedented efforts will be required in each of them.

For power generation, there are various means to reduce emissions of CO_2 from fossil fuels. Apart from simply using less fuel, the main options are to:

- switch to lower carbon fuels,
- develop higher-efficiency conversion processes, or
- apply CCS to power generation plants.

In the BLUE Map scenario, a significant contribution to the reductions in CO_2 emission required will come from power generation efficiency improvements and fuel switching (5%, or 2 Gt CO_2, in 2050). Improving the efficiency of power plants is important, and becomes more important the longer it takes for CCS to be deployed commercially at large scale. Higher-efficiency plants offer the multiple benefits of producing less CO_2 and other waste, and using less fuel and cooling water, per unit output of electricity. They are also more attractive for retrofitting capture technology, all other things being equal. At present, the average efficiency of the global fleet of coal-fired power plants is around 32% (LHV[1]). The existing fleet comprises mainly pulverised coal (PC) combustion plants. Best practice ultra-supercritical (USC) pulverised coal plants operate at an efficiency of around 45% (LHV).

Technologies have improved significantly over the years, particularly following improvements in materials that have permitted operation under supercritical (SC) and USC steam conditions, *i.e.* at temperatures well above 375°C and pressures higher than 22 megapascal (MPa). Figure 3 shows the percentage of coal-fired power generation capacity in various countries that employ the more efficient SC

[1] Efficiency reported on the basis of a fuel's lower heating value (LHV) and net electricity sent-out (net), i.e. LHV, net.

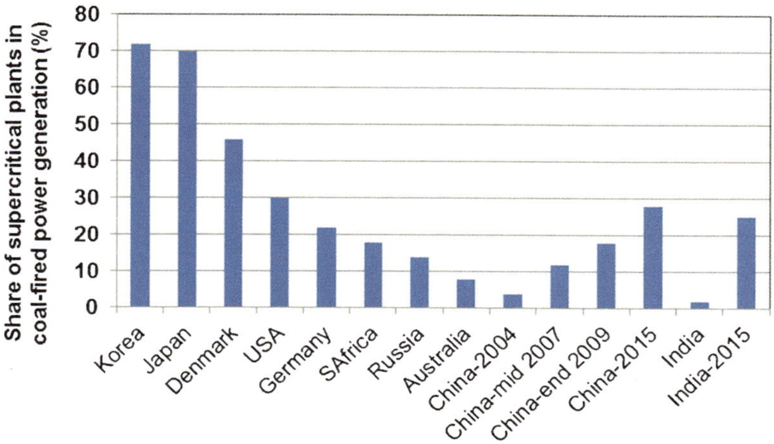

Fig. 3 Deployment of high-efficiency coal-fired power plants

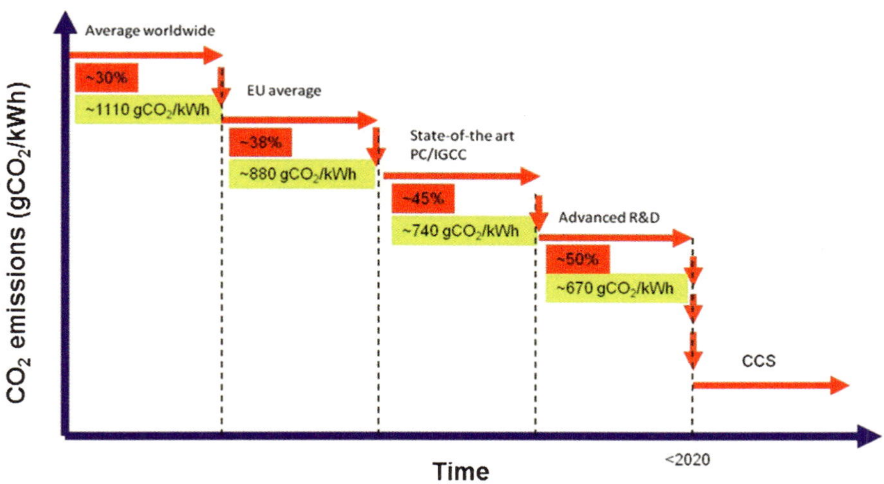

Fig. 4 CO_2 emission reductions through efficiency improvements at coal-fired power plants (indicated in *red*) (Source: Adapted from VGB 2009; Note: *PC* pulverised coal, *IGCC* integrated gasification combined cycle, *CCS* carbon capture and storage)

technology. From this figure, the relatively high efficiency of plants in Korea and Japan is clear. It can also be seen that China has increased considerably its deployment of supercritical technology. Before long, it is anticipated that the efficiency of coal-fired plants in China will approach the OECD average.

So, what can be done to raise the efficiency of coal-fired power plants from the present low global average of about 32% (LHV)? Figure 4 illustrates the advances that can be made. Within the European Union (EU), average efficiency is higher, at about 38% (LHV) and state-of-the-art plants around 45% (LHV). With advanced R&D, materials are being developed that will be capable of operating at even higher steam conditions. Using the new super alloys under development, steam conditions in excess of 30 MPa and 700°C could give rise to efficiencies of 50% or higher. But even then, emissions of CO_2 are still significantly higher than those from state-of-the-art natural gas combined cycle power plants, which typically emit around 350 g CO_2 per kWh.

The effective decarbonisation of power generation will require deep cuts in CO_2 emissions, much greater than those that can be achieved through increased generation efficiency or switching to natural gas. However, fossil fuels, and coal in particular, are expected to remain important fuels, especially in emerging economies. For China, India and Indonesia, for example, coal is and will remain the 'fuel of choice' for the foreseeable future – a low-cost, indigenous resource that offers energy security. Thus, CCS is an essential component of an effective, least-cost, low-carbon future. This is underscored by the IEA analysis that shows that, without CCS in the technology mix, the additional investment cost in the power sector to halve emissions of CO_2 from 2005 levels by 2050 would be significantly higher.

3 Carbon Capture and Storage

Almost one-fifth of the CO_2 reductions targeted in the BLUE Map scenario are projected to come from the deployment of CCS in 2050, with around half applied to power generation plants. So what exactly is carbon capture and storage?

In reality, CCS is not a single technology, but comprises three separate but integrated operations – CO_2 capture, CO_2 transport and CO_2 storage. Capture of the CO_2 is preferably from any large point source, of which a power plant is a good example. Following compression to a supercritical fluid, it is transported, typically via pipeline, to a storage site. There, it is injected into geological formations selected such that virtually all of the CO_2 will be permanently trapped in the subsurface. In this context, 'permanent' means that the injected CO_2 will be safely and securely stored over geological time scales, *i.e.* very long periods of time by any human measure – not tens or hundreds of years, but tens of thousands of years or more. The types of formation chosen for geological storage include saline aquifers (*i.e.* porous and permeable bodies of rock containing very salty water), depleted oil and gas reservoirs, or oil reservoirs that still contain mobile oil that may be displaced by injection of CO_2 and then recovered. Saline aquifers, although the least well characterised at present, appear to offer the greatest potential for storing CO_2 in the long run (Fig. 5).

As shown in Fig. 6, there are three main capture processes for power generation plants: post-combustion, in which CO_2 is separated from the flue gas following combustion; pre-combustion, in which CO_2 is separated from the fuel following gasification; and, finally, oxyfuel, in which fuel is burnt in oxygen rather than air, so that the combustion products are principally CO_2 and water, making the separation of CO_2 a much less onerous task.

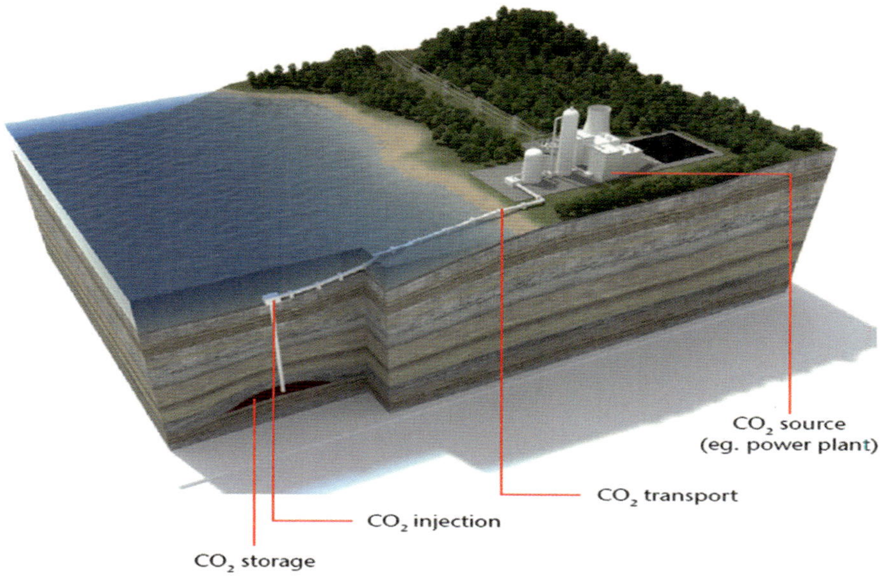

Fig. 5 Carbon capture and storage (Courtesy of Bellona Foundation)

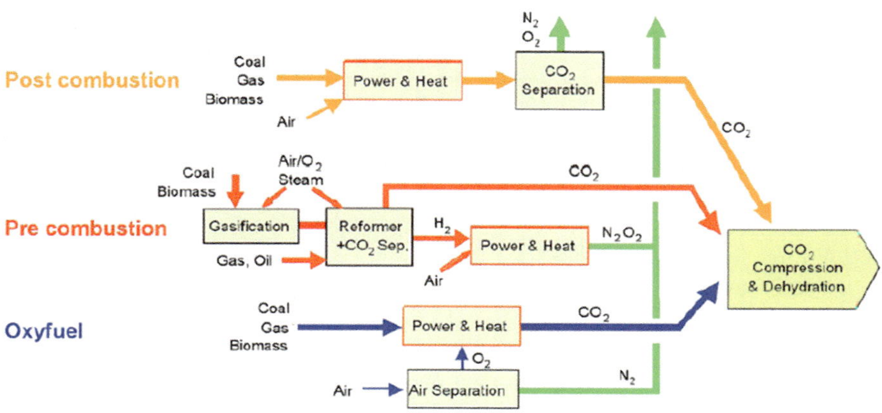

Fig. 6 CO_2 capture processes (Adapted from IPCC 2005)

Currently, there is no large-scale, integrated CCS demonstration on a power generation facility. In its survey of the status of CCS projects worldwide, the Global CCS Institute identified 77 currently active or planned large-scale, integrated projects, with around two-thirds of them in the power generation sector (GCCSI 2010). Most of these projects were located in developed countries, including Australia, Canada, Korea, the USA and nine European countries. It is notable,

Fig. 7 Vattenfall's 30 MW_{th} oxyfuel pilot plant (Germany) (Courtesy of Vattenfall)

however, that nine of the projects are in emerging economies, with five in China, three in the United Arab Emirates and one in Algeria. Three of the five planned Chinese projects were in the power sector.

Eight of the 77 large-scale integrated projects are in operation, although only five of them are operated for the purpose of permanent storage of CO_2. The other three are enhanced oil recovery (EOR) projects that, while they will undoubtedly store CO_2, do not presently undertake the monitoring necessary to verify that CO_2 will be permanently stored in these repositories.

In view of the potential of oxyfuel, both for high separation efficiencies and for retrofitting, significant activities are focused on testing the technology at the pilot scale. One example, shown in Fig. 7, is the Vattenfall oxyfuel carbon capture unit, a 30 MW_{th} demonstration plant being tested in Germany.

It is important to note that CCS is applicable not only to coal-fired plants and not only in the power sector. It can be employed on any large point source that emits CO_2, with half of the CO_2 captured in 2050 envisaged to come from the non-power sectors. In the power sector, CCS could be applied to coal-, gas- and biomass-fuelled power plants. In fact, its application to biomass is potentially important and opens the possibility of achieving 'negative CO_2 emissions' *i.e.* carbon from the atmosphere is fixed in biomass, which is then burnt, and the resulting CO_2 stored, which effectively results in the removal of CO_2 from the atmosphere. In industrial sectors such as iron and steel, cement, and pulp and paper, there are numerous opportunities to reduce CO_2 emissions to the atmosphere. In the upstream oil and gas sectors, gas processing has been recognised as one early opportunity; CCS has been applied on a commercial scale in this sector for many years.

Currently, more than 6 Mt of CO_2 are injected into geological formations for the purpose of storage each year. Though, in the short-to-medium term, the emphasis is on deploying CCS in the power sector, there is, as yet no large-scale power plant with fully integrated CCS.

Most CO_2 is likely to be transported by pipeline, and the United States is leading the way with nearly 6,000 km of CO_2 pipeline already installed and in service (Bliss et al. 2010). The US pipeline network transports nearly 60 Mt CO_2/year, most of which is used for enhanced oil recovery (Moritis 2009). It is also possible that CO_2 will be transported by sea tanker; although this option is obviously not applicable for overland transport, it would be attractive where captured CO_2 is to be transported over large distances to offshore storage sites.

4 CCS Challenges

Recognising the contribution of CCS to mitigating CO_2 emissions, the G8 leaders affirmed at their 2008 Summit in Hokkaido Toyako, Japan, that '*We strongly support the launching of 20 large-scale CCS demonstration projects globally by 2010, taking into account various national circumstances, with a view to beginning broad deployment of CCS by 2020*' (G8 2008). Since that statement was made, no new large-scale project has entered operation, but that is not to say that no progress has been made. Several governments have committed to provide over US$26 billion in funding support to facilitate the launch of at least 19 and, potentially, up to 43 large-scale CCS integrated demonstration projects by 2020 (IEA 2010c). This is very promising, as government support is vital in helping projects under development overcome the final hurdles. Nonetheless, substantial challenges lie ahead.

In October 2009, the IEA issued a CCS Technology Roadmap (IEA 2009) that identified the milestones to be achieved, and the major barriers, opportunities and policy measures to be met for CCS to make the contribution projected according to the BLUE Map scenario. The global technology deployment milestones presented in that Roadmap are shown in Fig. 8.

To meet the vision developed in the Roadmap, by 2020, 100 large-scale fully integrated CCS projects would be needed; by 2030, 850 projects; and by 2050, 3,400 projects. Each of these projects would need to store an average of 3 Mt CO_2/year. The scale of the challenge is clear. At present, as Fig. 9 shows, there are just five operating large-scale, integrated CCS projects – two in North America, two in Europe and one in Africa – each of which stores over 1 Mt CO_2/year, with the exception of the Norwegian Snohvit project, which stores around 700,000 t/year.

The relative contributions from the emerging and the more advanced economies are important. Although the OECD countries must take the lead in deploying CCS on a large scale, the Technology Roadmap suggests that, for CCS to meet its contribution, activities will need to be broadly the same in both OECD and non-OECD countries by 2020. This presents a huge challenge for advanced and emerging economies alike. To meet the challenge, it is recognised that progress also needs to

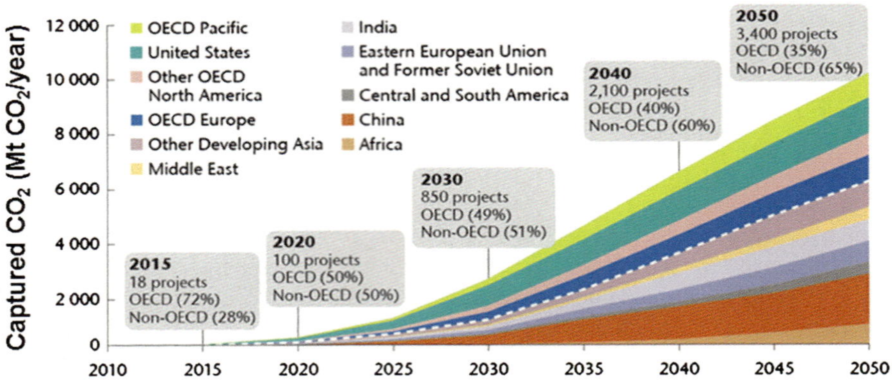

Fig. 8 The CCS Technology Roadmap: milestones in the global deployment of CCS by region (Source: IEA 2009)

be made in areas such as financing CCS and creating value for captured CO_2, in establishing suitable legal and regulatory frameworks, and in communication to promote public engagement.

It is important to note that China is already undertaking significant activities to develop CCS. Progress has been made on the GreenGen plan to construct a large-scale integrated gasification combined cycle (IGCC) plant with integrated CCS. The first phase of that plan – the IGCC plant with test facilities for the study of CO_2 capture – is now under construction. There is also much activity at the pilot scale. In June 2008, a small demonstration unit was completed at the Gaobeidian power plant in Beijing, capturing 3,000 t CO_2/year for use in the food and drinks sector (see Fig. 10). The results from that plant have subsequently led to the commissioning of a project in Shidongkou, Shanghai, which is capturing up to 100,000 t CO_2/year. Markets in the food and drinks sector have been found for the CO_2 captured from these plants.

There is also strong interest in other emerging economies, with work currently underway in South Africa.

A major barrier to the broad deployment of CCS, one that is often raised, is its cost and the associated efficiency penalty. Capture, transport and storage will all add to the cost of generating electricity. Of course, the lack of experience in building large-scale power plants with capture and in developing storage sites means that the magnitude of the cost has large uncertainties associated with it. The costs of capture are expected to be high, and to far exceed those of transport and storage.

Capture is expensive not only because of the cost of additional capture-related equipment, but also because it reduces the plant efficiency, which means that more gross generating capacity is needed for the same output and additional fuel must be consumed for each unit of electrical energy generated. A recent re-analysis of previously published studies by the IEA suggests that the average reduction in plant efficiency would be between 25% for a pulverised coal plant using post-combustion capture and 20% for an IGCC plant with pre-combustion capture (Finkenrath 2011).

Fig. 9 Approximate locations of operational CCS projects

Fig. 10 CO_2 capture unit at Huaneng Power's Gaobeidian plant, Beijing, China (Courtesy of China Huaneng Group)

Similarly, for a generating plant with the addition of capture, the overnight capital cost is estimated to increase by between 75% and 44% relative to the same plant without capture (Finkenrath 2011).

As a means to offset some of the additional costs needed to construct and deploy CCS, potential markets for the CO_2 captured are being sought. The term carbon capture, utilisation and storage (CCUS) is often used to describe this process. CCUS is, in fact, a sub-category of CCS. Its use in tertiary or enhanced oil recovery is a good example of a process where costs may be offset by utilisation of the captured CO_2. Though the potential to utilise a high proportion of the CO_2 captured in future is likely to be limited, it may well offer lower cost opportunities to deploy the technology in the short-to-medium term.

Where CCS is included on a new fossil fuel power generation facility, it is likely that the technology would be fully integrated into a highly efficient plant. Where it is retrofitted to an existing plant, it is likely that there would be a preference to maximise net output, *i.e.* for CCS to be retrofitted to a high-efficiency plant or to upgrade the plant to improve its base efficiency. In each case, net efficiency would be maximised, the increased fuel requirements would be minimised and CO_2 emissions would be low.

The additional costs of CO_2 transport and storage will be largely contained in the capital investment required, and will depend on a number of factors, including the distance between the CO_2 source and storage site; the amount of CO_2 to be handled; the type and nature of the transport infrastructure (*e.g.* shared versus dedicated pipelines); and the storage formation type, location and geological features. Progressive optimisation of pipeline and storage infrastructure over time is expected to bring about reductions in transport and storage costs.

CCS may be less expensive to apply in the industrial and upstream oil and gas sectors, where some attractive early opportunities for the technology may be found. For example, in natural gas processing, CO_2 must be separated from the gas extracted from high-CO_2 gas fields in order to make it a saleable commodity regardless of concerns about climate. Because the cost of separating CO_2 is a necessary part of producing such gas, the incremental cost of CCS can be low. This is the basis for four of the five large-scale, integrated CCS projects in operation today. In other industrial processes, such as cement production, CCS is likely to be more expensive to apply.

What has become clear is the importance of building new large-scale capture demonstration plants (Rubin et al. 2007). If, as forecast, the costs of capture technologies fall over time, the experience gained in operating these plants will be the primary factor. These plants will be where the major learning takes place. Operational experience is required to optimise capture technologies not previously applied at this scale. Demonstration plants will provide opportunities for developing and testing innovative solutions, and will help to build the necessary confidence in the technology among major stakeholders such as governments, industry, environmental NGOs and the public.

Where will the CO_2 be stored? Around the world there are areas where the prospects for their use as storage sites are high, and other large areas where they are very poor indeed. This is illustrated in Fig. 11.

National governments have begun to undertake initial, high-level estimates of their storage resources. The US Department of Energy, for example, estimates that the domestic storage resource, *i.e.* the amount of space available in geological formations that may be suitable for storage, is of the order of thousands to tens of thousands of gigatonnes of CO_2 in saline aquifers alone (NETL 2010). Similar work has also been undertaken in South Africa, where the storage capacity in saline aquifers has been estimated at approximately 150 Gt CO_2, most of which is located offshore (Council for Geoscience 2010). While it is certain that only a small fraction of storage resources will be technically suitable for storage and economically feasible to develop, they appear to be more than sufficient to handle the amount of CO_2 that could be captured in the near-term. The uneven distribution of storage resources, however, underscores the need to develop CO_2 transport infrastructure, and to ensure that international movements of CO_2 for storage are permitted under international law (IEA 2011).

For CO_2 to be stored safely and securely over geological time scales, a high level of confidence in prospective storage sites is required. To characterise a site to the extent required is both expensive and time consuming. To date, very little site-specific storage exploration has been undertaken. If the milestones proposed in the IEA's CCS Technology Roadmap are to be achieved, greater efforts are urgently required in this area (Friedmann et al. 2006).

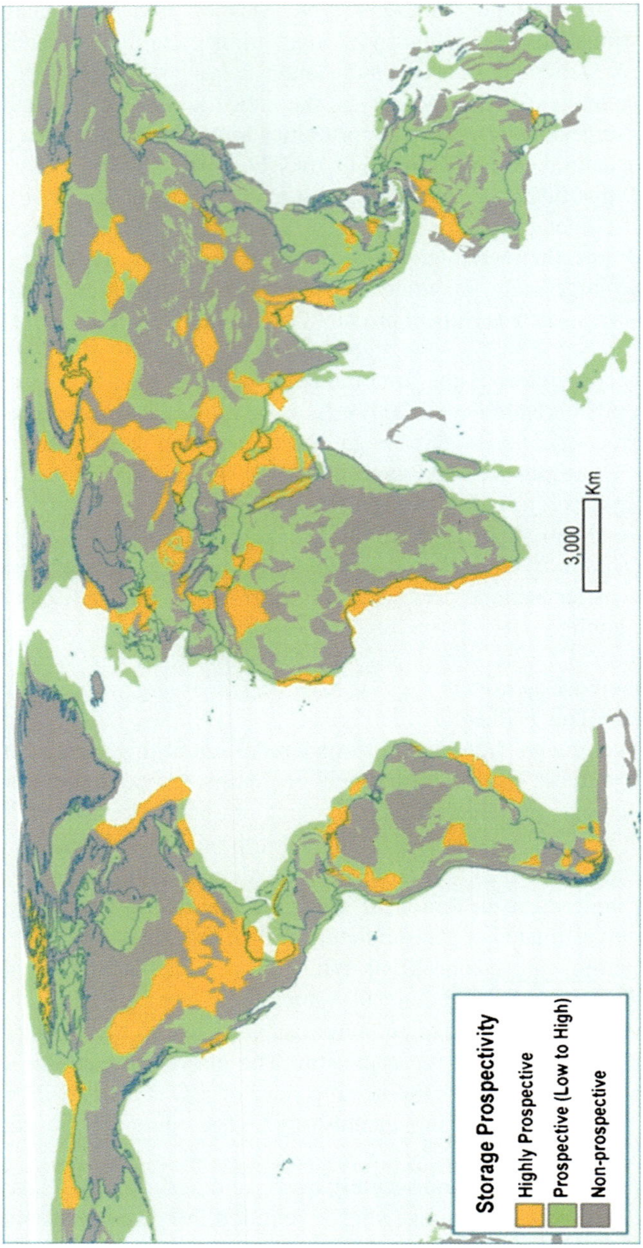

Fig. 11 Prospective sites for the storage of CO_2 (Adapted from IPCC 2005)

Accelerating the development of CCS technology will also require appropriate regulation to ensure that storage projects are developed in such a way that the local environment and human health and safety are protected; the stored CO_2 is permanently retained; and storage resources are developed in a responsible way that minimises conflicts with other users of the subsurface. Furthermore, mechanisms must be put in place that will allow operators and governments to manage effectively the long-term liabilities associated with storage sites (IEA 2010d). Governments have made progress in this area recently, although much remains to be accomplished. For example, in 2009 the EU issued a directive establishing a consistent European legal and regulatory framework for the geological storage of CO_2. By mid-2011, this directive had been transposed into legislation in many member states – a detailed discussion can be found in the IEA *CCS Legal and Regulatory Review* (IEA 2011). Other jurisdictions, most notably the Commonwealth of Australia, some Australian states, the United States and the Canadian province of Alberta, are well advanced in this area.

5 Summary and Conclusions

The rise in CO_2 emissions, including those from the energy sector, must be arrested and emissions must be reduced. Fossil fuels, and particularly coal, are the fuels of choice for new electricity generating plants in rapidly emerging economies such as China and India. This makes CCS an important component of any portfolio of measures to mitigate rising CO_2 emissions at least cost and to prevent the damaging consequences of climate change.

CCS comprises three separate but integrated operations, namely CO_2 capture, transport and storage. Each of these operations has been demonstrated at scale, but they have not yet been fully integrated into a power plant. Only a handful of large-scale, integrated CCS demonstration plants are in operation. Although other demonstration plants are at various stages of development, these and many more need to be operational over the next few years.

In its CCS Technology Roadmap, the IEA presents a set of requirements that the technology has to deliver in terms of CO_2 mitigation. The level of deployment required presents a tremendous global challenge. Meeting this challenge will depend on successful progress in many areas – technical, policy, legal, financial, market and organisational.

To successfully demonstrate the technology, finance is without doubt a critical factor. At present, CCS is expensive: it adds significantly to the capital and operating costs of the base plant to which it is applied. While further development and additional operating experience will reduce these costs, establishing a price for carbon and the expectation that this price will, on average, increase over time, are vitally important. It is also important that CCS is recognised in incentive mechanisms under post-Kyoto arrangements. Government support in the early stages will be crucial, as will the engagement of industry.

For CCS to be implemented effectively, regulation needs to be in place to reduce the risks. Derogations or exemptions may be possible for early demonstration plants, but broader deployment will require an effective legal and regulatory environment. Progress has been made, but much remains to be done.

Access to fossil fuels or to CO_2 storage capacity may influence local deployment. In countries with indigenous fossil fuel resources, CCS offers the potential to exploit them with minimal impact on the environment. Some countries may find that they have extensive potential storage resources, and this may well provide an incentive to encourage CCS, such as by developing a CO_2 transport infrastructure.

Another very important part of the equation is public engagement and acceptance of CCS technology. It has already been established that, in practice, without public acceptance, the technology will not progress at the rate required.

Substantial efforts are being made to develop CCS technology in both advanced and emerging economies. However, it is clear that if CCS is to play its full part in ensuring a secure, low-carbon energy future, the pace of progress will need to increase dramatically.

References

Bliss, K., Darrick, E., Harms, R. W., Carrillo, V. G., Coddington, K., et al. (2010). *A policy, legal, and regulatory evaluation of the feasibility of a national pipeline infrastructure for the transport and storage of carbon dioxide.* Oklahoma City: Interstate Oil and Gas Compact Commission. www.iogcc.state.ok.us/Websites/iogcc/Images/PTTF%20Final%20Report%202011.pdf

Council for Geoscience. (2010). *Atlas on geological storage of carbon dioxide in South Africa.* Johannesburg: South African Centre for CCS. www.sacccs.org.za/wp-content/uploads/2010/11/Atlas.pdf

Finkenrath, M. (2011). *Cost and performance of carbon dioxide capture from power generation* (IEA Energy Papers, No. 2011/05). Paris: International Energy Agency. www.oecd-ilibrary.org/energy/iea-energy-papers_20792581

Friedmann, S. J., Dooley, J. J., Held, H., & Edenhofer, O. (2006). The low cost of geological assessment for underground CO_2 storage: Policy and economic implications. *Energy Conversion and Management, 47*, 1894–1901.

G8 (Group of Eight). (2008, July 8). *G8 Hokkaido Toyako summit leaders' declaration*. Hokkaido Toyako. www.mofa.go.jp/policy/economy/summit/2008/doc/doc080714__en.html

G8 (Group of Eight). (2010, June 25–26). *G8 Muskoka declaration: Recovery and new beginnings.* Muskoka, Canada. http://g8.gc.ca/g8-summit/summit-documents/g8-muskoka-declaration-recovery-and-new-beginnings

GCCSI (Global CCS Institute). (2010). *The global status of CCS: 2010.* Canberra: Global CCS Institute. http://www.globalccsinstitute.com/publications/global-status-ccs-2010

IEA (International Energy Agency). (2009). *Technology roadmap: Carbon capture and storage.* Paris: OECD/IEA. www.iea.org/papers/2009/CCS_Roadmap.pdf

IEA (International Energy Agency). (2010a). *Energy technology perspectives 2010 – scenarios & strategies to 2050.* Paris: OECD/IEA. www.iea.org/w/bookshop/add.aspx?id=401

IEA (International Energy Agency). (2010b). *World energy outlook 2010.* Paris: OECD/IEA. http://www.iea.org/weo/2010.asp

IEA (International Energy Agency). (2010c). *Carbon capture and storage: Progress and next steps* (IEA/CSLF Report to the Muskoka 2010 G8 Summit). Paris: OECD/IEA. www.iea.org/papers/2010/ccs_g8.pdf

IEA (International Energy Agency). (2010d). *Carbon capture and storage: Model regulatory framework* (IEA Energy Papers, No. 2010/12). Paris: IEA. www.oecd-ilibrary.org/energy/iea-energy-papers_20792581

IEA (International Energy Agency). (2011). *Carbon capture and storage legal and regulatory review.* Paris: OECD/IEA. www.iea.org/ccs/legal/review.asp

IPCC (Intergovernmental Panel on Climate Change). (2005). *IPCC special report on carbon dioxide capture and storage.* Cambridge: Cambridge University Press. www.ipcc.ch/publications_and_data/publications_and_data_reports_carbon_dioxide.htm

IPCC (Intergovernmental Panel on Climate Change). (2007). *Climate change 2007* (The Fourth IPCC Assessment Report). Geneva: IPCC. www.ipcc.ch/publications_and_data/publications_and_data_reports.htm

Moritis, G. (2009). More CO_2–EOR projects likely as new CO_2 supply sources become available. *Oil and Gas Journal, 107*(45), 41–47.

NETL (National Energy Technology Laboratory). (2010). *Carbon sequestration atlas of the United States and Canada.* Pittsburgh: US Department of Energy. www.netl.doe.gov/technologies/carbon_seq/refshelf/atlasIII/2010atlasIII.pdf

Rubin, E. S., Yeh, S., Antes, M., Berkenpas, M., & Davison, J. (2007). Use of experience curves to estimate the future cost of power plants with CO_2 capture. *International Journal of Greenhouse Gas Control, 1*(2), 188–197.

VGB (Vereinigung der Großkesselbesitzer eV). (2009). *Facts and figures: Electricity generation.* Essen: VGB PowerTech. www.vgb.org/daten_stromerzeugung-dfid-25744.html

Nuclear Energy

Jacques Bouchard

Abstract The growing global population and the expanding world economy need increasing amounts of energy at reasonable prices. In addition to fossil fuels, other energy sources will be needed, including nuclear and renewables. Today, nuclear energy provides about 13% of the electric power produced worldwide. The installed nuclear capacity of 366 GW could be multiplied by a factor of 3 or 4, reaching 1,200–1,500 GW by the middle of the century. Progress has been made in improving the safety, reliability and economic performance of generation III reactors now being built. Generation IV systems, to be deployed around the middle of the century, should ensure safe, secure and sustainable energy production for a longer time horizon.

1 Energy Challenges for the Twenty-first Century

This chapter addresses some of the complex issues surrounding nuclear energy. It assesses the current status of the nuclear industry and what the future holds, assuming that energy consumption continues to grow (IEA 2010). There are geopolitical tensions, climate risks and other complicating issues. The climate challenge, discussed in earlier chapters (Schelling 2012; Burnard and McCoy 2012), leads to the difficult question of how to reduce CO_2 emissions while producing more energy. There is no one solution, but rather several partial solutions, all of which have to be considered if the world is to succeed in ensuring sustainable energy supplies.

Worldwide energy needs will increase for many reasons, as discussed in earlier chapters (e.g. Pachauri 2012; Luna and Gomelsky 2012; Difiglio 2012). While there

J. Bouchard (✉)
Atomic Energy Commission (CEA), Batiment 121, 91191 Gif-sur-Yvette Cedex, France
e-mail: jacques.bouchard@cea.fr

F.L. Toth (ed.), *Energy for Development: Resources, Technologies, Environment*, Environment & Policy 54, DOI 10.1007/978-94-007-4162-1_16,

are abundant fossil resources in some places, it is clear that oil and gas are becoming more expensive if not scarcer, although recent advances in shale gas exploration and extraction are modifying this picture somewhat (IEA 2011). Climate change due to CO_2 emissions is a real concern. Nuclear energy is an important part of the solution, complementing alternative solutions discussed in other chapters: energy efficiency (Kan 2012), renewable energy sources (Gupta 2012) and coal- and gas-fired power plants with carbon capture and storage (CCS) (Burnard and McCoy 2012).

As of September 2011, more than 430 power reactors were operating in 30 countries, with a total installed capacity of 366 GW. Those plants generated 13% of the electric power produced worldwide, and 6% of the primary energy – approximately the same amounts as for hydropower. The installed nuclear capacity could be increased by a factor of 3 or 4, reaching 1,200–1,500 GW by the middle of the century. This is the level in most reliable forecasts (Toth 2008; IAEA 2010; IEA 2010). This expansion will be possible with the existing light-water reactors that are the most common types in operation around the world. The countries that will build such reactors in the coming decades aim to operate them for at least 60 years, and so will have to take into consideration uranium supply issues even if there is no difficulty at first. There is also a need for a clear vision of waste management. This is certainly one of the issues on which the nuclear industry must clarify its position in the near future.

2 Nuclear Energy Development

What are the key requirements for nuclear energy in the twenty-first century? This section addresses safety, economy, waste management, uranium resources and proliferation resistance.

Figure 1 presents the current classification of nuclear power reactors. The first-generation reactors (first builds), are already history, while the second-generation plants represent the current industrial capacity for electricity production. The third generation of advanced reactors are what we are able to build now. Nuclear industry vendors around the world are proposing new designs, some of which are already under construction. Generation IV reactors are now being developed as systems for the future. We are looking to the wider use of nuclear energy worldwide, although some major improvements, in particular to ensure the sustainability of nuclear energy, will be needed.

2.1 Safety

Compared with other energy sources, nuclear is reliable and safe, even though the risk of a nuclear accident can never be completely eliminated. The data and records built up over the last 20 years have shown improvements in many areas, and users

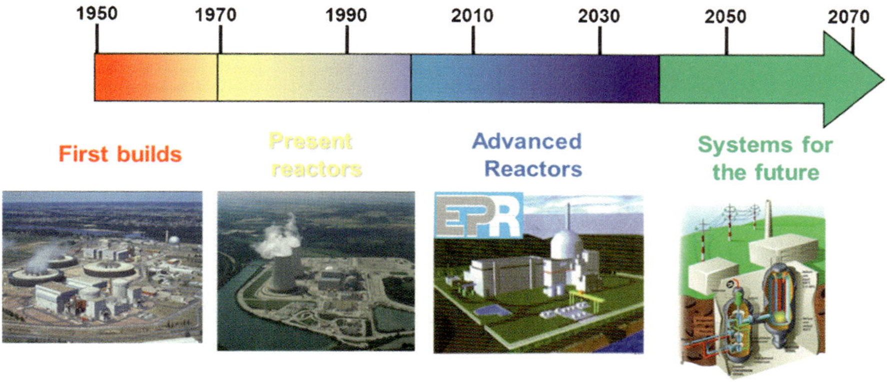

Fig. 1 Four generations of nuclear reactors (Note: *EPR* European pressurised water reactor)

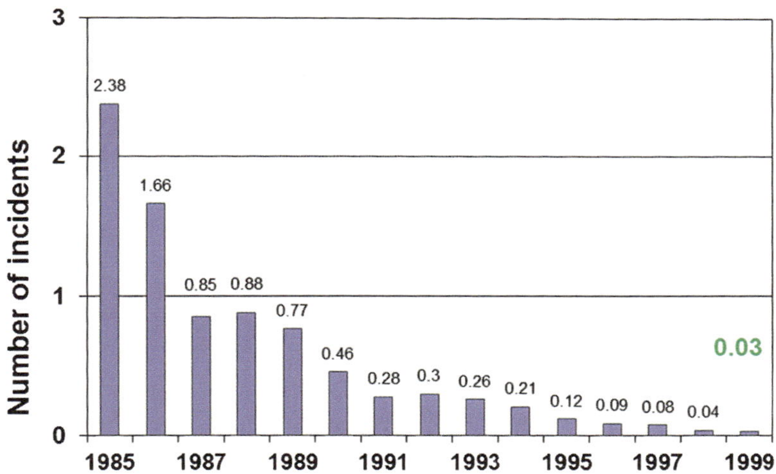

Fig. 2 Safety performance of generation II reactors

are generally satisfied by the safety of their plants. The number of incidents at existing plants, i.e. generation II reactors has declined (see Fig. 2), and further progress will be achieved with generation III plants, where improving safety is a permanent objective.

This improving safety record was interrupted in March 2011 by the incident at the Fukushima Daiichi nuclear power plant in Japan, when a rare combination of two powerful natural disasters – a earthquake of magnitude 9.0 on the Richter scale, followed by a 15 m high tsunami – led to severe damage of the plant and releases of radioactivity to the atmosphere and the ocean. Even if the health consequences are very limited, the severity of the incident in an area already devastated by the earthquake and tsunami prompted a strong emotional response around the world.

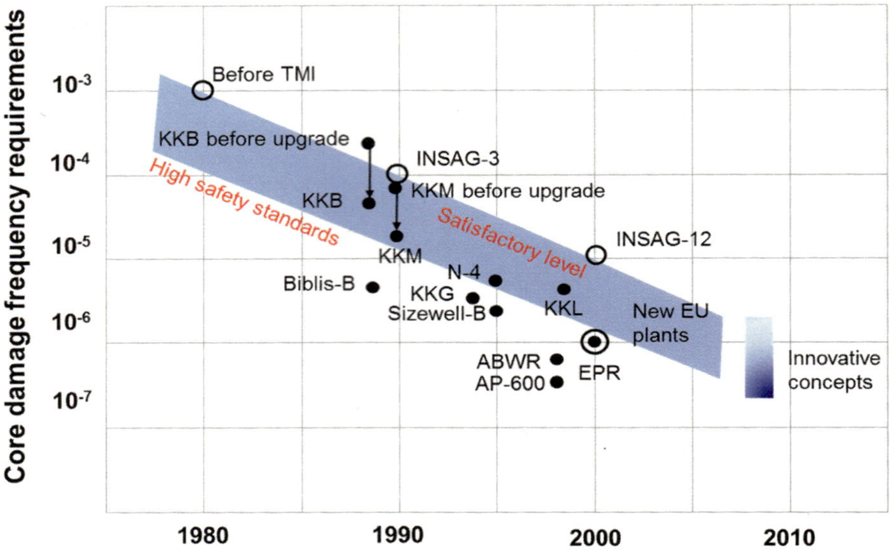

Fig. 3 Evolution of safety requirements to reduce the risk of core damage (Notes: *TMI* Three Mile Island, *KKB* nuclear power plant Brunsbüttel, *KKM* Mühleberg nuclear power plant, *INSAG* International Nuclear Safety Group (see IAEA 1988, 1999), *N-4* new four-loop reactor (named after the four-loop nuclear steam supply system in the French 1,450 MWe plant), *EPR* European Pressurised Water Reactor, *Biblis-B* Nuclear power plant Biblis/Rhine, Unit B, *ABWR* advanced boiling water reactor, *AP-600* 600 MWe advanced passive plant, *KKG* Gösgen nuclear power plant, *KKL* Leibstadt nuclear power plant)

All the countries involved in nuclear energy production immediately decided to proceed with a thorough check of their plants' ability to withstand such rare and extreme natural events and, more generally, their ability to deal with extreme situations such as a complete power blackout or the loss of cooling water supply systems.

The lessons learned from the Fukushima event will help to make existing and future reactors safer. A better way to manage the risk of severe accidents was already the basis for generation III designs. Once a complete analysis of the Fukushima incident has been made, these designs will be improved, if needed, and greater attention will be paid to characterising proposed plant locations before siting decisions are made.

Figure 3 shows that in Europe the core damage frequency requirements, considered the most severe accident at a nuclear plant, have declined steadily in recent years. Note that in the figure the scale is logarithmic, so it is clear that real progress has been made. Nevertheless, in the design and construction of new reactors there are still opportunities to reduce this risk even further.

The design of generation III advanced reactors has benefited from experiences at existing plants, and has taken into account the lessons of past events, in particular the Three Mile Island accident in the United States in 1979. In this new generation of reactors, the main aim has been to improve safety while preserving economic

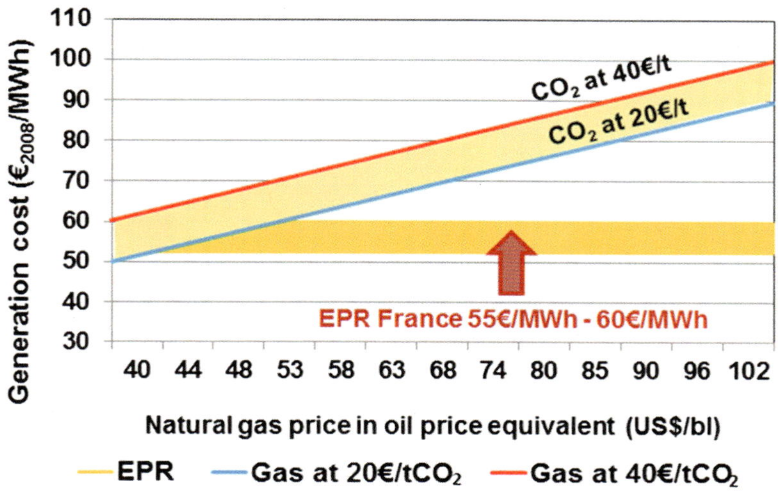

Fig. 4 Projected generation costs for 2015 baseload production: nuclear and gas combined cycle as a function of gas prices at two levels of CO_2 emission penalties (Note: *EPR* European Pressurised Reactor)

competitiveness. Many studies have been made, and the products being proposed today show a large potential for further safety improvements. Several approaches have been tested, such as building smaller or larger reactors, and including passive or active safety systems. There are many technical solutions, and all of them have led to real progress in mitigating the consequences of potential large accidents. Despite the negative impacts of the Fukushima accident, safety remains a strong point in favour of nuclear energy.

2.2 *Economics*

From the economic point of view, at least in some countries, nuclear energy is very competitive. In Europe this is clearly the case; all the studies carried out over the last decade have shown that nuclear energy is competitive (IEA/NEA 2010). It will become even more competitive if the prices of oil and gas continue to rise, and if progressive carbon taxes are introduced in some countries (see Fig. 4).

The global competitiveness of nuclear energy today is not an issue. The real issue is investment, because the capital costs are particularly high. The fuel cost is low, and operation and maintenance (O&M) costs are similar to those of fossil fuel plants of the same size. The capital costs of existing nuclear plants are higher than for most other energy sources, except for some renewables, and are probably similar to the cost of coal-fired plants when carbon capture technology is included (see Fig. 5).

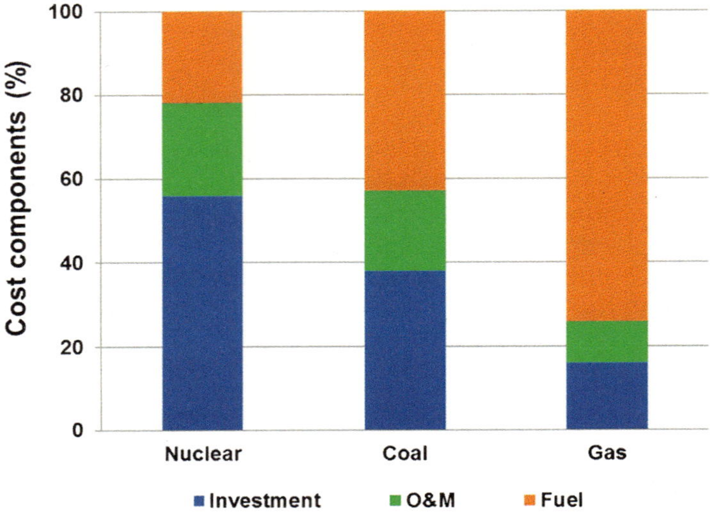

Fig. 5 Shares of cost components in three main power generation technologies (Note: *O&M* operation and maintenance)

The capital investment is needed for a period of approximately 10 years, without any return, making it difficult for some countries to implement nuclear energy on a large scale. In the future the nuclear industry must reduce the capital costs in comparison with its competitors.

2.3 *Waste Management*

Nuclear waste management is complicated because there are several possible solutions. Although the waste issue is being debated everywhere, it should be remembered that the quantities of radioactive waste are small. The big advantage of nuclear energy is its efficiency. In terms of the quantities of materials involved, the nuclear reaction is a million times more efficient than most chemical reactions, and so produces very little waste. But however much is produced, this waste still has to be managed. There are three options at the back end of the fuel cycle (see Table 1).

The way we treat the spent fuel from reactors is a key issue for waste management. Either we opt for an open cycle, in which all the spent fuel is disposed of as waste. This has been seen as an important option, as it was assumed to be cheaper and limited the immediate risk of proliferation (Forsythe 2009; Net 2010). Or we opt for a closed cycle which means recycling everything that can be recycled. This entails reprocessing the fuel and treating as waste only that part that cannot be burned in the reactor (Kim et al. 2009; Collins and Ehinger 2010; Zrodnikov 2010; Žagar et al. 2011). Most countries are now convinced that sooner or later a closed cycle is essential for sustainability.

Table 1 Back end of the fuel cycle: three options

	Open cycle	Closed cycle	Interim storage of spent nuclear fuel
Pros	Assumed to be cheaper	A sustainable option	'Wait & see' policy Easy for immediate decision
	Limits the immediate risk of proliferation	Reduces the volume and radiotoxicity of waste	Limits the short-term costs
	No immediate release of radioactivity	Industrial experience with reprocessing and recycling	Could be easier for public acceptance
Cons	An unsustainable option	Assumed to be more expensive	'Wait & see' policy A burden for next generation
	Large quantities of long-lived elements in the waste	Needs improved processes to limit the risk of proliferation	Everything remains to be done in the future
	Increased long-term risk of proliferation (Pu mine)	Large plants necessary for economy	The total cost will be much higher

Note: *Pu* plutonium

In addition to these two solutions there is a third option – wait and see. This means putting all nuclear waste in storage and waiting before deciding on the final solution. Adopting a wait and see policy has short-term advantages and long-term drawbacks. Storing spent fuels today is the cheapest way to proceed, but it is not a final solution. Storage duration will be limited for technological reasons, and in the end it will still be necessary to choose between disposing of the spent fuel and recycling. Delaying this final choice means transferring the burden to future generations and creating technical and financial problems that should have been resolved by those who now benefit from nuclear energy production.

Ultimately, all of these solutions will need geological disposal. Even with a closed cycle, even if we burn all that we can burn, and recycle all that can be recycled, at the end we still need geological disposal sites for the final waste. There has been a lot of work on the geological disposal of radioactive waste and most of the countries operating nuclear plants today have a project of this kind. Some have already established underground laboratories to examine practical issues such as geological conditions, and they are well advanced. Finland and Sweden, for example, are preparing to open repositories, and France is about to decide where to open a geological repository by 2025. The technical problems have already been mastered, but political and public acceptance issues remain to be resolved.

Progress has also been made in minimising the amounts of radioactive waste, such as by burning more and recycling in a closed fuel cycle. Figure 6 shows three waste management strategies: disposal of spent fuel without reprocessing, plutonium recycling, and the partitioning and transmutation of minor actinides. If we recycle only what is interesting from an economic point of view, the result will be close to the middle curve in Fig. 6. But if we burn all the actinides that can be burned, that will move us to the lower blue curve. In this case, the radiotoxicity of the waste

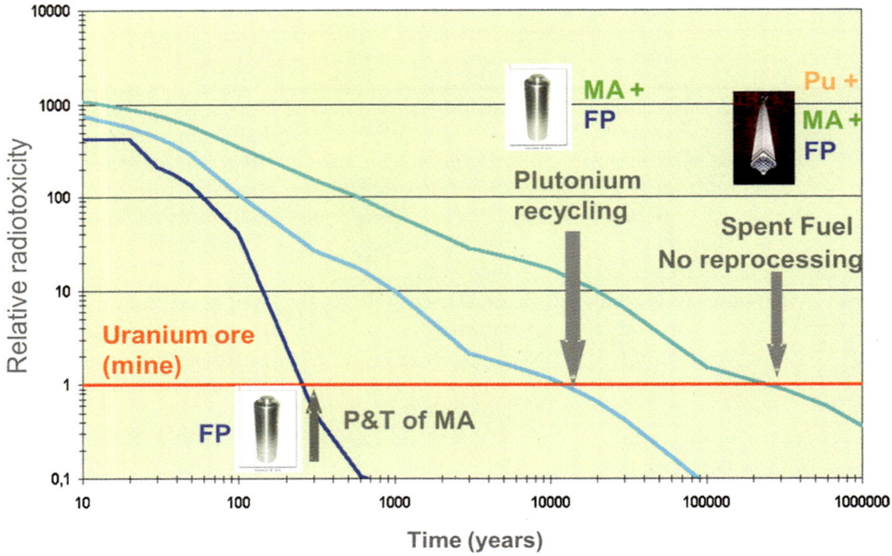

Fig. 6 Radioactive waste management strategies (Notes: *MA* minor actinides, *FP* fission products, *P&T* partitioning and transmutation, *Pu* plutonium)

decreases very quickly over time. Even if we still need to put radioactive waste in geological repositories, both the toxicity and risk are reduced in the future. So for waste too we have a solution. Today, all waste is stored provisionally without any risk. This means that we have the possibility of opening quite soon the geological disposal sites that are needed, but we have still the possibility of reducing waste volume by reprocessing.

2.4 Uranium Resources

Many reports have confirmed that uranium resources are sufficient to supply existing reactors in the short term (Mian et al. 2009; NEA and IAEA 2010; von Hippel 2010; Kahouli 2011). Based on what is known about estimated or speculative resources, we can operate all the reactors that are expected to be built over the next 50 years to the end of the century (see Fig. 7). However, if we stay with the existing light-water reactors, uranium consumption will increase. Looking at all the uranium needed for the reactors that will be constructed, by 2060 we will have already consumed all the speculative resources, as well as those estimated. Beyond 2060, the nuclear industry must move to phosphates, and maybe to sea water, as sources of uranium (see Rogner 2012).

There are several other possibilities, but they will become increasingly expensive. This is one of the reasons why work continues on sustainability, and on a new type

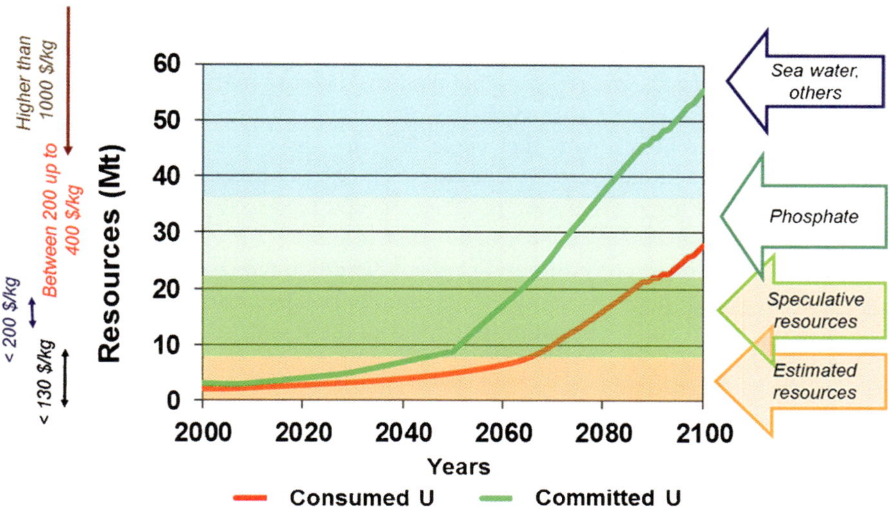

Fig. 7 Uranium sources and costs (Note: The *two curves* indicate uranium use assuming that global nuclear capacities continue to increase to the end of the twenty-first century. The 'consumed U' curve (*red*) indicates the amount of uranium effectively consumed by a given date (e.g. the *value* indicated for 2080 is the total amount of uranium consumed from the beginning to 2080). The 'committed U' curve (*green*) indicates the total amount of uranium that will be consumed at the end of life of all the plants in operation at a given date (e.g. the *value* indicated for 2080 is the sum of the uranium consumed by this date and the uranium that will be consumed by all the plants in operation in 2080 until their end of life))

of reactor, the fast neutron reactor, which will allow us to recycle all that can be recycled of the spent fuel. In this case we will reduce the need for uranium for power reactors by a factor of 50–100, so there will be no more problem in the future. That is why we can say that uranium resources can allow practically indefinite power supplies as long as the spent fuel is recycled in fast reactors.

2.5 Proliferation Resistance

Proliferation resistance is the "characteristic of a nuclear energy system that impedes the diversion or undeclared production of nuclear material or misuse of technology by States in order to acquire nuclear weapons or other nuclear explosive devices" (IAEA 2002). Ensuring the proliferation resistance of nuclear energy systems is critically important, and raises a number of issues (Chirayath et al. 2008; Feiveson et al. 2008; Carlson 2009).

First, from a technical point of view, the light-water reactors currently in use do not pose a problem. They use low enriched uranium, a fuel that cannot be used for anything else. They produce electricity, and it is very difficult to use these reactors for other purposes without being clearly identified. The problem is the fuel cycle associated with these reactors. They require enrichment plants to produce the enriched uranium fuel, and reprocessing plants if a closed fuel cycle is chosen, and there are risks associated with both the plants and the transportation of fissile materials. The problems with fuel cycle activities have been clearly identified, although these are limited to a few countries. There is a limited risk, but clearly the difficulty of ensuring proliferation resistance is confined to the fuel cycle, and not the reactors themselves, at least for light-water reactors.

Second, the problem of proliferation is not directly linked to energy production. Even if we halt all nuclear energy production the risk of proliferation will remain because there will still be uranium in the world, there will still be knowledge and there will be technological progress. So we cannot resolve the problem completely by considering energy production alone. Nevertheless, progress can still be made. In the development of generation IV nuclear energy systems, it is particularly important that efforts are made to reduce the risk of proliferation from the technical point of view.

There are many other aspects of proliferation resistance already considered and measures are taken around the world to achieve an ever-improving control of nuclear material through the safeguards systems managed by the IAEA. In addition the nuclear industry is working on other possibilities of technical progress to reduce the proliferation risk.

3 Prospects for Nuclear Energy

In the coming decades, a number of new countries, including developing nations, will implement their own nuclear energy programmes. Nuclear energy is a structural choice for a country. It does not involve simply ordering a reactor and then letting a company operate it. Once a state has decided to pursue nuclear energy, it needs to establish strong organisations that will be responsible for safety, waste management and many other functions. The magnitude of this process should never be underestimated.

Today, energy demand is rising, and the nuclear industry is ready to satisfy that demand. The industry went through some difficult times about 10 years ago because no new plants were ordered. But it has survived and is now developing new products, including generation III reactors. There is no problem with the supply of materials. Obviously, if the global nuclear energy production capacity is increased, it will be necessary to increase industrial capacity and the number of engineers working in the field.

Climate change demands that we increase reliance on nuclear energy, not only in countries that are already well equipped and are prepared to expand their capacity, but also in new countries, particularly those with rapidly growing economies. Looking at the countries that are already nuclear, most of them are major emitters of CO_2, particularly the United States, Europe, Russia and China. Indeed, all of the countries that could increase their nuclear capacity are the most important emitters of CO_2. Thus, there is an urgent need to extend the use of nuclear energy in these countries. By the middle of the century, these countries will represent more than 80% of global nuclear capacity, so that is where most of the associated problems will remain.

For newcomer countries, nuclear energy could promote economic development while limiting the use of fossil fuels. But for those countries that choose to make use of this energy source, several requirements will need to be considered. The first is the development of relevant skills and appropriate infrastructure. A lot of work in these areas is underway, with support from the IAEA. The second, most important requirement is to establish a robust organisation responsible for nuclear safety. Safety regulations must be clearly formulated. Whatever activities a country is willing to subcontract to foreign companies, it is essential to have an organisation to execute the country's responsibilities, in particular for safety regulation. The third step involves the choice of the reactor that corresponds to the need of a robust financial engineering to build a project successfully. To develop the skills, infrastructure and safety systems, partnerships with countries with experience in operating nuclear plants would offer many benefits.

The IAEA has published a series of documents detailing the required process – what should be done, and in what order, in which areas and when – to enable a country to develop nuclear energy (IAEA 2007a, b). The benefits of nuclear energy production make it an attractive option, and should be available to all countries that wish to use it.

4 Conclusions

A rapidly developing world with a growing population needs large amounts of energy at reasonable prices. The supplies of fossil fuels that have dominated the energy market for nearly two centuries cannot continue forever. It will be increasingly necessary to rely on complementary energy sources, including nuclear and renewables.

Today, nuclear energy is reliable, safe and less expensive than most other sources of energy. Progress has been made in improving performance, and the generation III reactors now under construction should be able to meet the increasing demand for energy in the next few decades. The generation IV systems, which will be deployed around the middle of the century, should ensure secure and sustainable energy production for a longer time horizon.

Besides the optimistic nuclear capacity forecasts in countries that already rely on this energy source, there is a growing interest in newcomer countries. The nuclear industry is ready to answer the demand, but these countries should give priority to developing the necessary skills and infrastructures, and in particular safety and security systems, as recommended by IAEA.

References

Burnard, K., & McCoy, S. (2012). Fossil fuels and carbon capture and storage. In F. L. Toth (Ed.), *Energy for development: Resources, technologies, environment* (pp. 187–203). Dordrecht: Springer.

Carlson, J. (2009). *Introduction to the concept of proliferation resistance* (ICNND Research Paper No. 8, Revised). Barton: Australian Safeguards and Non-Proliferation Office.

Chirayath, S., Metcalf, R., Ragusa, J., & Nelson, P. (2008, March 30–April 4). *Assessment of proliferation resistance requirements for fast-reactor fuel-cycle facilities*. In 8th International Conference on Facility Operations–Safeguards Interface, Portland, OR.

Collins, E. D., & Ehinger, M. H. (2010). Identification of adequate proliferation resistance for a closed fuel cycle. *Transactions of the American Nuclear Society, 102*, 126–127.

Difiglio, C. (2012). Financing power sector investments. In F. L. Toth (Ed.), *Energy for development: Resources, technologies, environment* (pp. 161–170). Dordrecht: Springer.

Feiveson, H., Glaser, A., Miller, M., & Scheinman, L. (2008). *Can future nuclear power be made proliferation resistant?* College Park: Center for International and Security Studies at Maryland.

Forsythe, J. (2009). *3 R's of nuclear power: Reading, recycling, and reprocessing: Making a better tomorrow for Little Joe*. Bloomington: Author House.

Gupta, S. (2012). Financing renewable energy. In F. L. Toth (Ed.), *Energy for development: Resources, technologies, environment* (pp. 171–186). Dordrecht: Springer.

IAEA (International Atomic Energy Agency). (1988). *Basic safety principles for nuclear power plants* (INSAG Safety Series No. 75-INSAG-3). Vienna: IAEA.

IAEA (International Atomic Energy Agency). (1999). *The safe management of sources of radiation: Principles and strategies* (INSAG Safety Series No. 11). Vienna: IAEA.

IAEA (International Atomic Energy Agency). (2002). *Proliferation resistance fundamentals for future nuclear energy systems*. Vienna: IAEA.

IAEA (International Atomic Energy Agency). (2007a). *Considerations to launch a nuclear power programme*. Vienna: IAEA.

IAEA (International Atomic Energy Agency). (2007b). *Milestones in the development of a national infrastructure for nuclear power*. Vienna: IAEA.

IAEA (International Atomic Energy Agency). (2010). *Energy, electricity and nuclear power estimates for the period up to 2050*. Vienna: IAEA.

IEA (International Energy Agency). (2010). *World energy outlook 2010*. Paris: IEA.

IEA (International Energy Agency). (2011). *Are we entering a golden age of gas? Special report*. Paris: IEA.

IEA, & NEA (International Energy Agency, & Nuclear Energy Agency). (2010). *Projected costs of generating electricity*. Paris: IEA/NEA.

Kahouli, S. (2011). Re-examining uranium supply and demand: New insights. *Energy Policy, 39*(1), 358–376.

Kan, C.-C. F. (2012). Energy efficiency for development. In F. L. Toth (Ed.), *Energy for development: Resources, technologies, environment* (pp. 113–120). Dordrecht: Springer.

Kim, H. J., Jun, E., Chang, S. H., & Kim, W. H. (2009). An assessment of the effectiveness of fuel cycle technologies for the national energy security enhancement in the electricity sector. *Annals of Nuclear Energy, 26*(5), 604–611.

Luna, N., & Gomelsky, R. (2012). Development, infrastructure and energy: Exploring the linkages in Latin America. In F. L. Toth (Ed.), *Energy for development: Resources, technologies, environment* (pp. 95–112). Dordrecht: Springer.

Mian, Z., Nayyar, A. H., & Rajaraman, R. (2009). Exploring uranium resource constraints on fissile material production in Pakistan. *Science and Global Security, 17*, 77–108.

NEA, & IAEA (Nuclear Energy Agency, & International Atomic Energy Agency). (2010). *Uranium 2009: Resources, production and demand*. Paris/Vienna: NEA/IAEA.

Net, T. (Ed.). (2010). *Energy: Nuclear*. Alexandria: Capitol Net Inc.

Pachauri, S. (2012). Demography, urbanisation and energy demand. In F. L. Toth (Ed.), *Energy for development: Resources, technologies, environment* (pp. 81–94). Dordrecht: Springer.
Rogner, H.-H. (2012). Energy resources. In F. L. Toth (Ed.), *Energy for development: Resources, technologies, environment* (pp. 149–160). Dordrecht: Springer.
Schelling, T. C. (2012). Rethinking energy aid mechanisms: Three premises. In F. L. Toth (Ed.), *Energy for development: Resources, technologies, environment* (pp. 27–34). Dordrecht: Springer.
Toth, F. L. (2008). Prospects for nuclear energy in the 21st century: A world tour. *International Journal of Global Energy Issues, 30*, 3–27.
von Hippel, F. (Ed.). (2010). *The uncertain future of nuclear energy* (Research Report 9). Princeton: International Panel on Fissile Materials. www.fissilematerials.org/ipfm/site_down/rr09.pdf
Žagar, T., Buršič, A., Špiler, J., Kim, D., Chiguer, M., et al. (2011). Recycling as an option of used nuclear fuel management strategy. *Nuclear Engineering and Design, 241*(4), 1238–1242. doi: 10.1016/j.nucengdes.2010.04.007 www.sciencedirect.com/science/article/pii/S0029549310002608 - aff1.
Zrodnikov, A. V. (2010). Fast reactors in the energy security for the stable development of Russia. *Atomic Energy, 108*(4), 230–233.

Energy and Food Security

Faris Hasan

Abstract This chapter explores the complex linkages between energy and food security. The production of biofuels has increased significantly in view of their supposed benefits in terms of mitigating climate change and improving energy security. Concerted efforts are being made to develop new forms of biofuels, and to increase the production of biofuel crops over vast areas. Given the current state of the technology, the intended increase raises concerns about their economic and ecological viability. The chapter demonstrates that when setting biofuel targets, most governments do not consider the full range of implications. The benefits of biofuels include a 3–8% increase in agricultural value added in the developed countries and a reduction in annual greenhouse gas emissions of 12.4 Gt CO_2 equivalent by 2050. But the costs will be substantial: the loss of 30–45 million hectares of arable land to biofuel production and a significant increase in the number of people at risk of hunger.

1 Background

Over the last decade the production of biofuels has increased significantly in view of their supposed benefits in terms of mitigating climate change and improving energy security by reducing the dependency on fossil fuels. Concerted efforts are being made worldwide to develop new forms of biofuels, and to increase the production of biofuel crops over vast areas. These efforts, however, are raising genuine

F. Hasan (✉)
OPEC Fund for International Development (OFID), P.O. Box 995, A-1011 Vienna, Austria
e-mail: info@ofid.org

F.L. Toth (ed.), *Energy for Development: Resources, Technologies, Environment*, Environment & Policy 54, DOI 10.1007/978-94-007-4162-1_17,

concerns about the economic and ecological viability of biofuels in their current form, and whether future gains will justify the necessary investments. There are many examples of successes in the development of biofuels, but the scale of the envisaged increase in production in many countries raises concerns about their economic and environmental impacts in the coming decades. The critical question is simply whether the benefits outweigh the costs. The costs are and will remain significant, especially in view of the large areas of land required and the direct competition between biofuels and traditional agricultural crops.

While there are many linkages between energy and food security, this chapter is concerned with just two issues: the effects of increased biofuel production on food prices, and whether biofuels could threaten food security. The latter question stems from practical concerns about the possible impacts of biofuels, as illustrated in the following statement from the World Bank: '*we estimate that a doubling of food prices over the last three years could potentially push 100 million people in low income countries deeper into poverty.*' (World Bank 2008, p. 3).

The OPEC Fund for International Development (OFID) has worked on poverty alleviation since the 1970s and has spent more than US$10 billion tackling the problems associated with poverty. For OFID, it is vitally important to clarify this matter. OFID believes in progress and the development of renewable sources of energy, but not at any cost, especially if they affect those most in need.

2 Biofuel Scenario Assumptions

To ensure a considered and unbiased approach to the issue of biofuels, OFID decided to commission the International Institute for Applied Systems Analysis (IIASA) to carry out a study of biofuels and fuel security (OFID and IIASA 2009). In particular, IIASA was asked to look into the implications of accelerated biofuel production. Starting from a relatively low base, the levels of production have increased significantly between 2004 and 2008 (see Fig. 1); ethanol production increased by 300%, while that of biodiesel surged by 500%. This spectacular growth, according to proponents, will help to mitigate climate change, enhance energy security and foster rural development. In practice, however, the picture is far more complex and many important questions remain open.

The rise in biofuel production has been achieved at significant cost and is by no means sustainable. Many developed and a few developing countries have introduced mandatory, voluntary or indicative targets for the use of biofuels, together with considerable support in the form of subsidies, so the high rate of growth in biofuel production is hardly surprising (see Table 1). It should also be noted that most of the current mandatory targets extend up to 2020. It is this drive to continue raising production that is cause for concern, and prompted OFID to look into its implications. In recent years, the prices of fossil fuels have risen significantly, although in some cases, they are no higher in real terms than they were in the 1980s. So the argument that biofuels are a cost-effective alternative to fossils fuels is at best

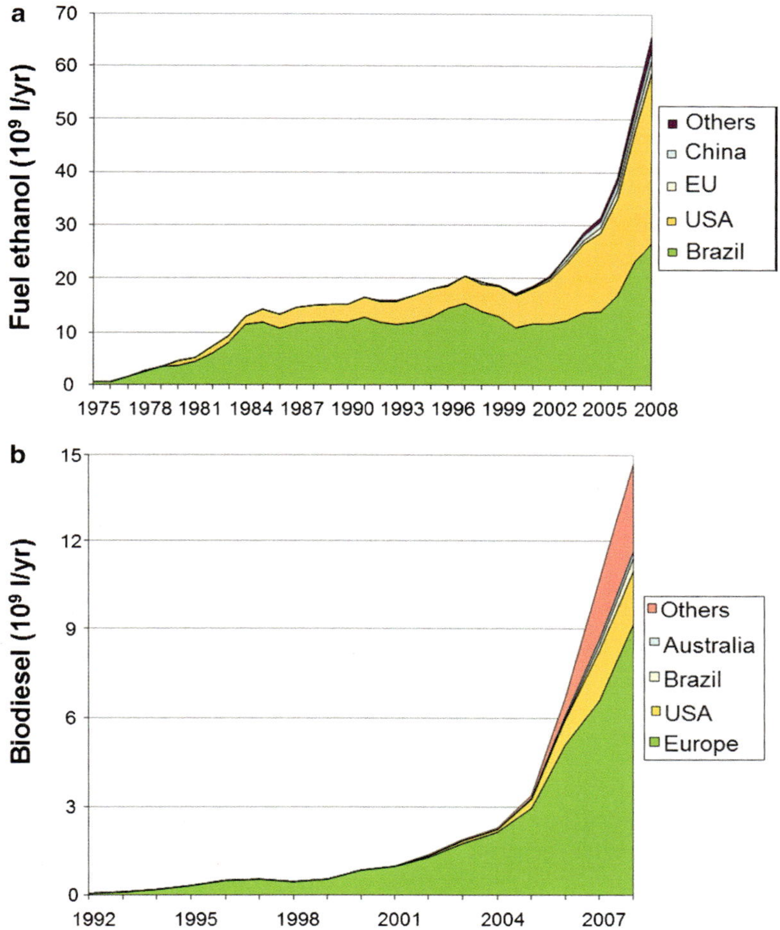

Fig. 1 Biofuels: ethanol production 1975–2007, and biodiesel production 1992–2007 (Sources: Licht 2007, 2008). (**a**) Fuel ethanol production 1975–2007. (**b**) Biodiesel production 1992–2007

misleading, and both financial and environmental implications must be questioned and carefully analysed.

Although nine scenarios were developed, defined and analysed, only two are discussed in this chapter in detail. Both scenarios take into account the difference between first- and second-generation biofuels. The former are produced mostly from food crops: oil crops (rapeseed, sunflower, soybean, oil palm, jatropha), sugar crops (sugarcane, sugar beet, sweet sorghum) and starch crops (wheat, rye, triticale, maize, sorghum, cassava). Second-generation biofuels, on the other hand, are produced from ligno-cellulosic materials, including herbaceous plants (Miscanthus, switchgrass, reed canary grass) and woody plants (poplar, willow, eucalyptus) and algae, all of which are non-food plants.

Table 1 Mandatory, voluntary or indicative targets for the use of biofuels in various countries

Country/region	Mandatory, voluntary or indicative targets
Australia	At least 350 million litres of biofuels by 2010
Brazil	Mandatory targets: 25% ethanol and 5% biodiesel blended with gasoline by 2010
Canada	5% renewable content in gasoline by 2010
China	2 million tonnes (Mt) ethanol by 2010, increasing to 10 Mt by 2020; 0.2 Mt biodiesel by 2010, increasing to 2 Mt by 2020
European Union	5.75% by 2010, 10% by 2020
France	7% by 2010, 10% by 2015, 10% by 2020
Germany	6.25% by 2010 10% by 2020
India	5% ethanol blended in gasoline in 2008, 10% as of 2009
	Indicative targets: 20% ethanol and 20% biodiesel blended with gasoline by 2017
Indonesia	2% biofuels in the energy mix by 2010, 3% by 2015, and 5% by 2020
Japan	0.6% of automobile fuel by 2010; a goal to reduce dependence on oil in the transportation sector from 98% to 80% by 2030
New Zealand	3.4% target for both gasoline and diesel by 2012
South Africa	2% biofuels by 2013
Thailand	2% biodiesel blended with gasoline by 2008, and 10% biodiesel and 10% ethanol by 2012
United States of America	12 billion gallons[a] by 2010, rising to 20.5 billion gallons by 2015 and to 36 billion gallons by 2022 (with 16 billion gallons from advanced cellulosic ethanol)

[a]1 US gallon = 3.785 l

The first scenario, TAR-V1, examines the transportation fuel demand as projected by the International Energy Agency (IEA 2008), assuming that the current biofuel targets are implemented and that second-generation biofuels are gradually deployed after 2015. The second scenario, TAR-V3, includes the previous assumptions as well as the accelerated development of novel conversion technologies and/or use of second-generation biofuels.

The study used the data presented in the *World Energy Outlook 2008* (IEA 2008) and the biofuel targets as inputs for IIASA's system of models, which includes the Agro-ecological Zone and World Food System models. These models have been used by the UN Food and Agriculture Organization (FAO) and have been peer-reviewed by academics. They have also been used in other studies, including those by the Intergovernmental Panel on Climate Change (IPCC) and the World Summit on Sustainable Development. With these models it is possible to assess the effects of biofuel targets on greenhouse gas (GHG) emissions, food security and rural development, and other factors such as the conversion of arable land and deforestation. Finally, the models can also be used to estimate the impacts of biofuels on energy security.

The models use a development scenario that is based on the results of a survey of relevant issues: population growth, economic development and energy demand. These results are fed into a climate model and then into the Agro-ecological Zones model, which in turn takes into account the impacts of climate change on crops. The next step is to input these data into the World Food System model whereby the

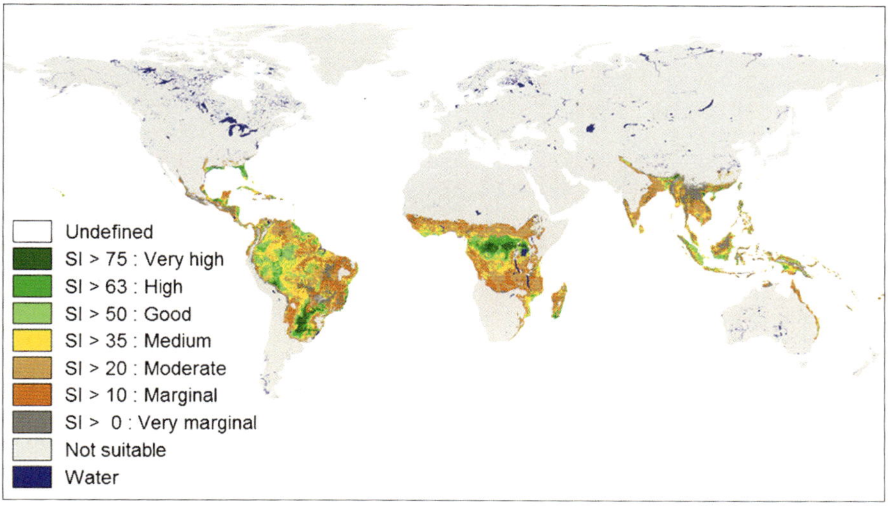

Fig. 2 Global distribution of land suitable for rainfed sugarcane production, high input level (Source: OFID and IIASA 2009; Note: *SI* suitability index)

production, demand, supply and the surpluses of each country are transferred via trade to the world market. The Agro-ecological Zones model calculates the suitability of different world regions for growing the various biofuel crops. For example, Fig. 2 illustrates the areas suitable for rainfed sugarcane production, taking into consideration the high water inputs needed. The two regions most suitable for sugarcane production are South America and Central Africa. However, in both regions, especially the latter, large areas of forest would need to be cleared for cultivation.

Apart from sugarcane, jatropha is now also widely cultivated for its seeds, from which oil is extracted for use as a fuel. Although the toxicity and yields of jatropha are still under discussion, the areas where it could be grown are potentially very large. Many areas of Africa, for example, are better suited to growing jatropha than sugarcane, and are not concentrated in the centre of the continent (see Fig. 3). However, the large volumes of water required for jatropha cultivation will continue to restrict the areas where it is grown and will add significantly to the costs of production. Estimates of jatropha seed yields vary widely due to the genetic diversity of the species, the broad range of environments in which it is grown and jatropha's perennial life cycle. Jatropha seed yields can range from 1,500 to 2,000 kg per hectare, from which about 540–680 l of oil can be extracted. In short, yields will vary according to the quality of the land in which it is grown, just like any other plant.

However, it is important to remember that jatropha is not a food crop, and so in theory it should have little impact on food supplies. This highlights the major difference between first- and second-generation biofuels, and the increasing preference for the latter worldwide (see Fig. 4). Second-generation biofuel crops are far less restricted in terms of where they can be grown, and can for example also be planted in the United States. As a result, their potential benefits are great as long as the right

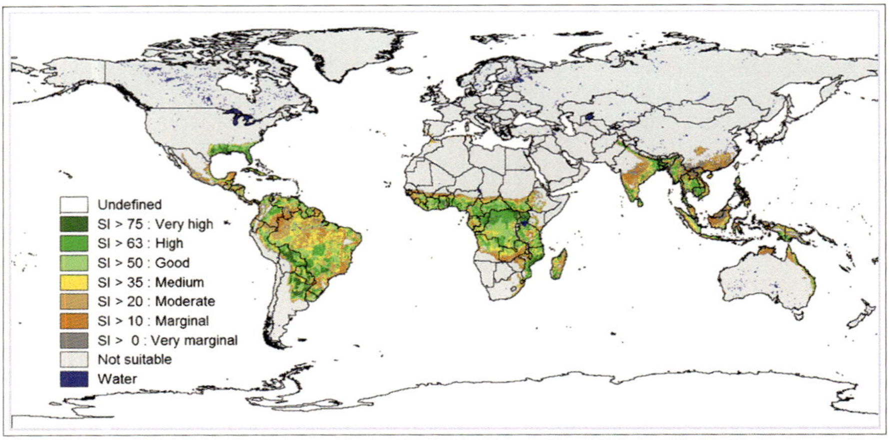

Fig. 3 Global distribution of land suitable for rainfed jatropha production (Source: OFID and IIASA 2009; Note: *SI* suitability index)

conditions for their cultivation, especially adequate water supplies, are available. Jatropha can also be intercropped with other cash crops such as coffee, sugar, fruits and vegetables.

3 Biofuel Scenario Results

It is widely assumed that increased use of biofuels will help to curb GHG emissions and thus mitigate the impacts of climate change. To investigate this assumption, we used the IIASA models to estimate the net GHG emissions savings resulting from the use of biofuels under the various scenarios, as explained above (see Fig. 5). According to the models, the use of biofuels will help to reduce some GHG emissions. However, the changes in land use resulting from the conversion to biofuel crops will have a negative impact on GHG emissions. The net effect is decidedly negative for both scenarios until 2020, and GHG emissions actually rise as a result of increasing biofuel production and use. The TAR-V1 scenario gives the greatest negative savings (i.e. emissions increases), and TAR-V3 somewhat lower savings. The lowest negative GHG emissions savings are obtained for the scenario without accelerated biofuel targets. This third scenario corresponds to the projection of the *World Energy Outlook 2008* (IEA 2008), labelled WEO in Fig. 5.

By 2030, the scenarios show substantial improvements in the net effects of the production and use of biofuels, with GHG emissions savings becoming positive for both the TAR-V3 and WEO scenarios. However, if first-generation biofuels continue to be used, as in TAR-V1, GHG emissions savings will be negative until 2030. For all three scenarios, positive savings will begin only in 2050, i.e. with a 40-year time lag.

a

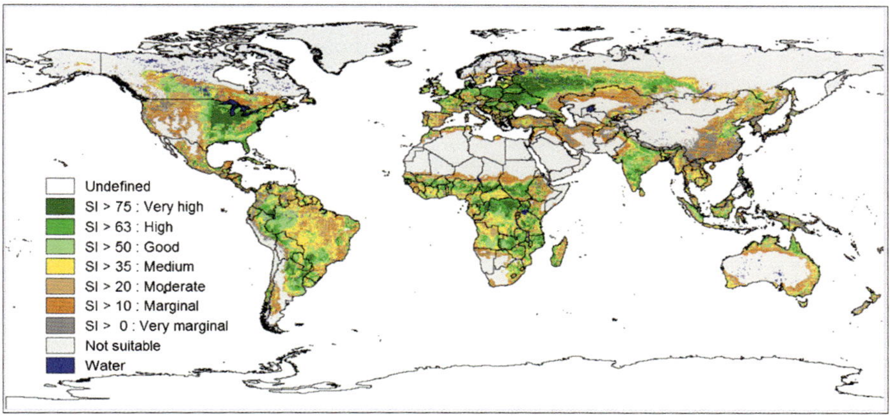

b

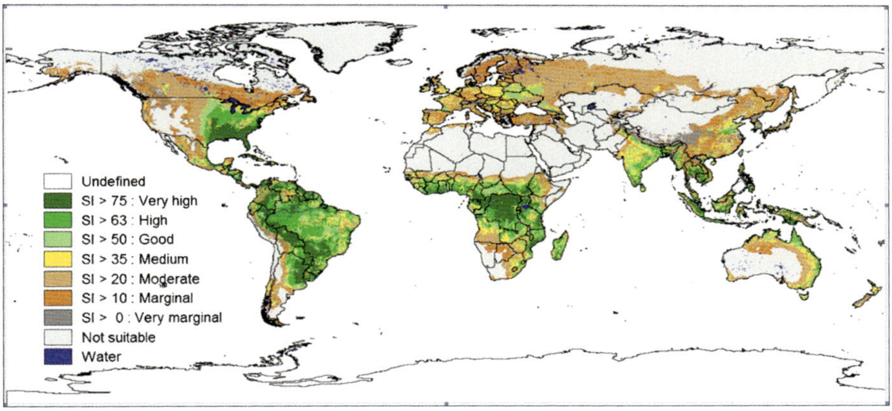

Fig. 4 Global distribution of land suitable for producing first- and second-generation biofuel crops (Source: OFID and IIASA 2009). (**a**) First-generation biofuels. (**b**) Second-generation biofuels. Note: *SI* suitability index

The implications for the environment aside, food prices were considered relative to a reference scenario in which biofuel production is kept at the 2008 level (see Fig. 6). There is only a moderate increase in food prices in the WEO scenario, whereas prices surge in TAR-V1 and TAR-V3. Under the TAR-V1 scenario, which assumes an accelerated use of first-generation biofuels and the introduction of second-generation biofuels commencing as early as 2015, cereal prices increase by 35% and those of other food crops by almost 30%. Under the TAR-V3 scenario, the price increases are slightly lower. Yet even this forecast is doubtful because second-generation biofuels are still in the research phase, and the assumption that they will be deployed by 2015 is very optimistic. If they were to be deployed before 2015, however, then the impact on agricultural prices is less severe.

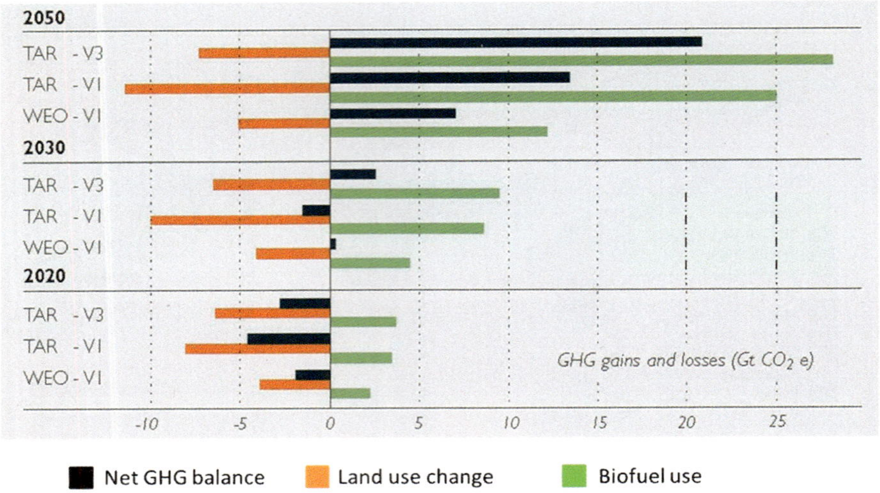

Fig. 5 Net greenhouse gas (*GHG*) emissions savings achieved in selected scenarios (Source: OFID and IIASA 2009; Note: Computations for first-generation biofuels are based on greenhouse gas saving coefficients (EC 2008) and the IPCC's tier 1 approach for carbon losses due to land use changes (IPCC 2006). For second-generation biofuels a greenhouse gas saving of 85% was used)

According to the World Food System model, the prices of coarse grains such as corn and wheat will increase the most. However, animal feedstock prices will fall because of the abundant supplies of protein 'cakes' produced as a by-product in the manufacture of biofuels that can be fed to animals. Biofuels are expected to increase value added and bring developmental value for the rural sector in developing countries. The model also calculates that the increase in agricultural value added is higher for the developed than for developing countries in each scenario (see Fig. 7). Moreover, even though agricultural value added is higher, the largest increase is only 5% in 2020 and 8% in 2030, which casts doubt on the idea that biofuels will promote development by increasing agricultural output.

The higher demand for biofuels will result in increased cereal production, but the increase will be higher in the developed than in developing countries (see Fig. 8). The reason for this, as we have seen in past crises, is that developing countries respond to price increases much more slowly than developed countries. As a result, the increase in cereal production is modest, reaching a maximum of 250 million tonnes in 2030.

The disproportionate rise in the output of biofuels relative to the increase in cereal production raises the question of where the cereals used to make biofuels will come from. In 2020, 66% will come from the increased cereal production of about 150 million tonnes. As a result, cereal prices will rise and the demand for food and animal feed will fall. However, the decrease in demand for animal feed will be in developed countries while the decrease in demand for food in developing countries (see Fig. 9).

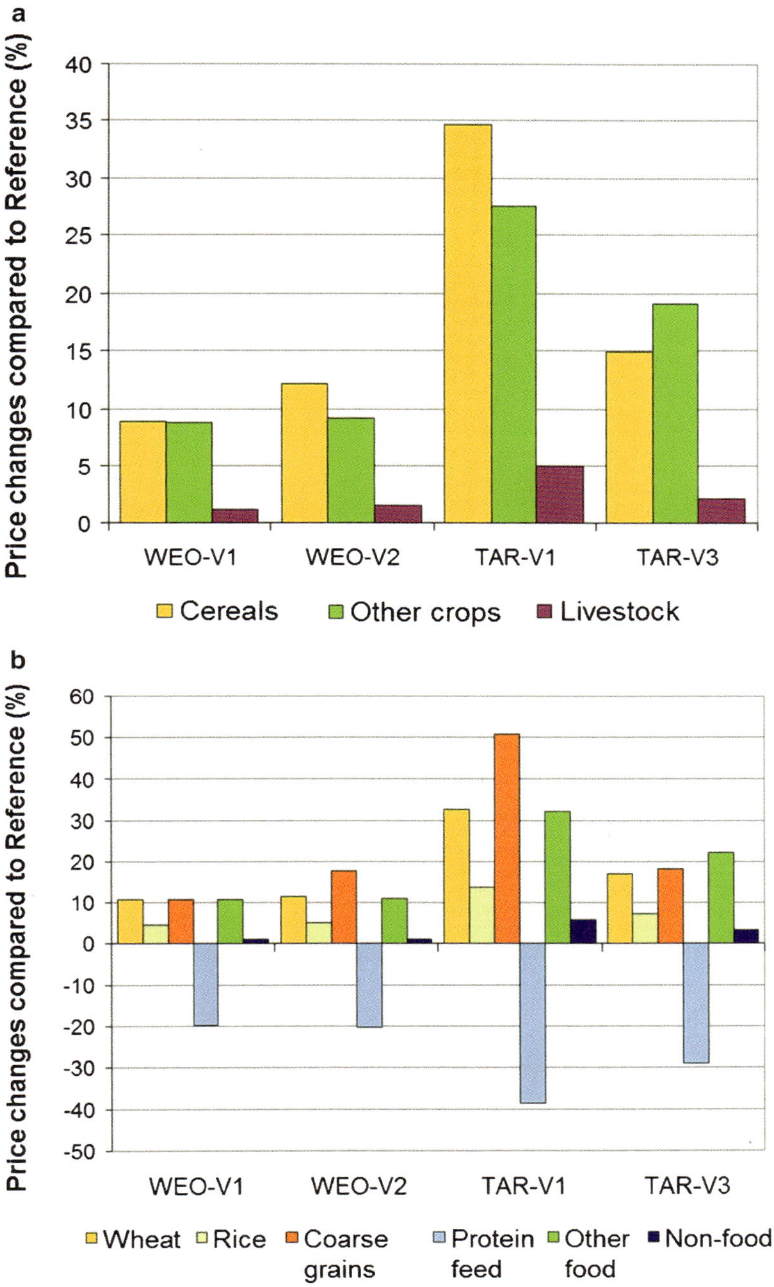

Fig. 6 Impact of first-generation biofuels on the prices of agricultural products and selected commodities in 2020. Price changes compared with the reference scenario (Source: OFID and IIASA 2009). (**a**) Main product groups. (**b**) Selected commodities and aggregates

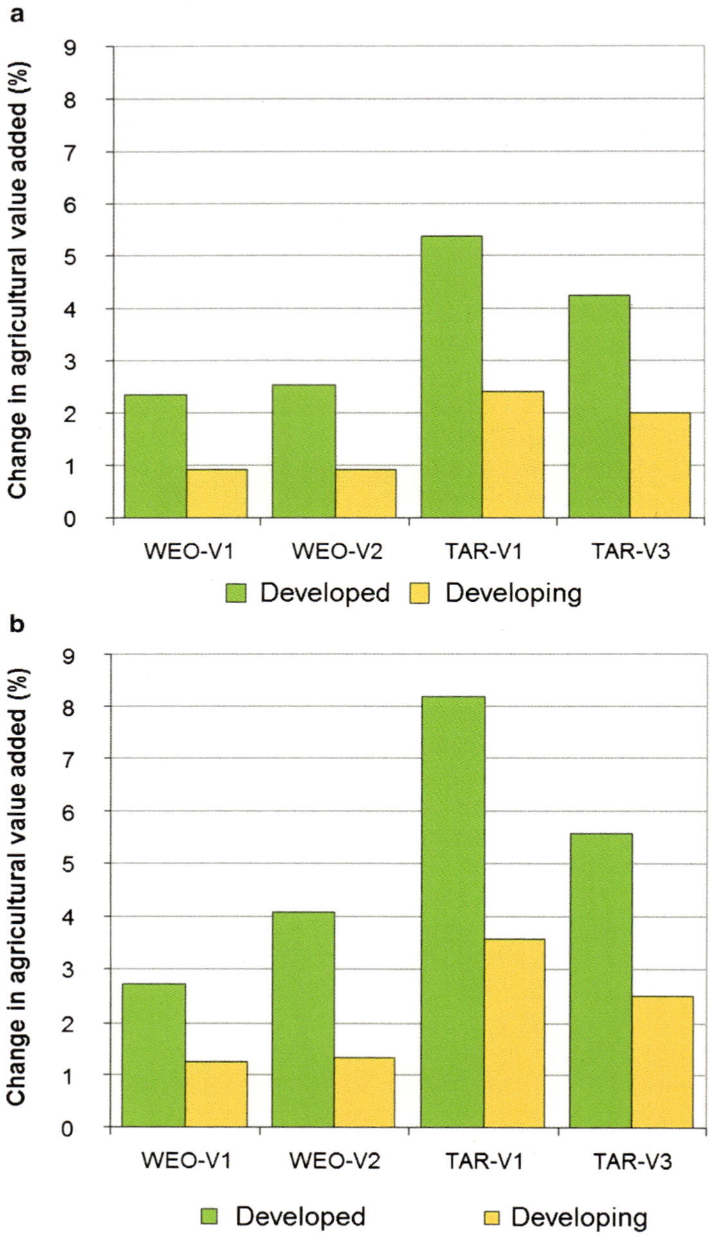

Fig. 7 Impact of biofuels on agricultural value added. Changes relative to the reference scenario (Source: OFID and IIASA 2009). (**a**) 2020. (**b**) 2030

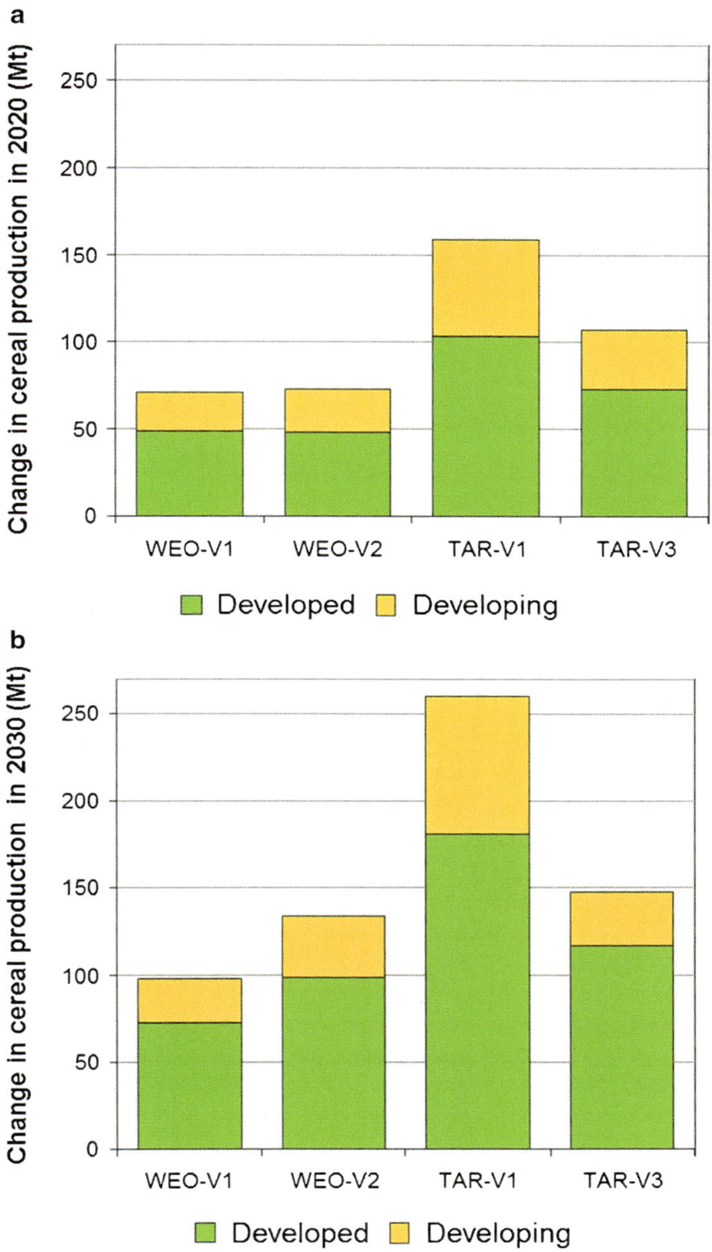

Fig. 8 Changes in cereal production in 2020 and 2030 relative to the reference scenario (Source: OFID and IIASA 2009). (**a**) 2020. (**b**) 2030

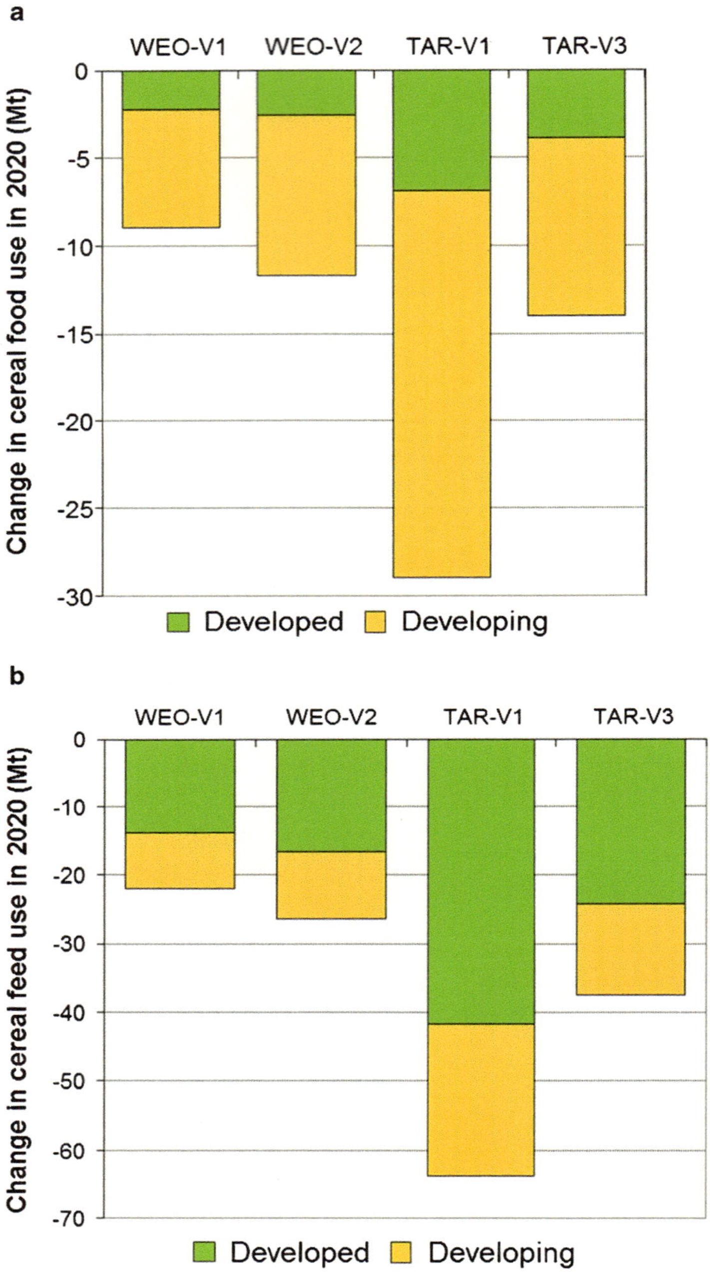

Fig. 9 Changes in the consumption of cereals and animal feed relative to the reference scenario (Source: OFID and IIASA 2009). (**a**) Food. (**b**) Animal feed

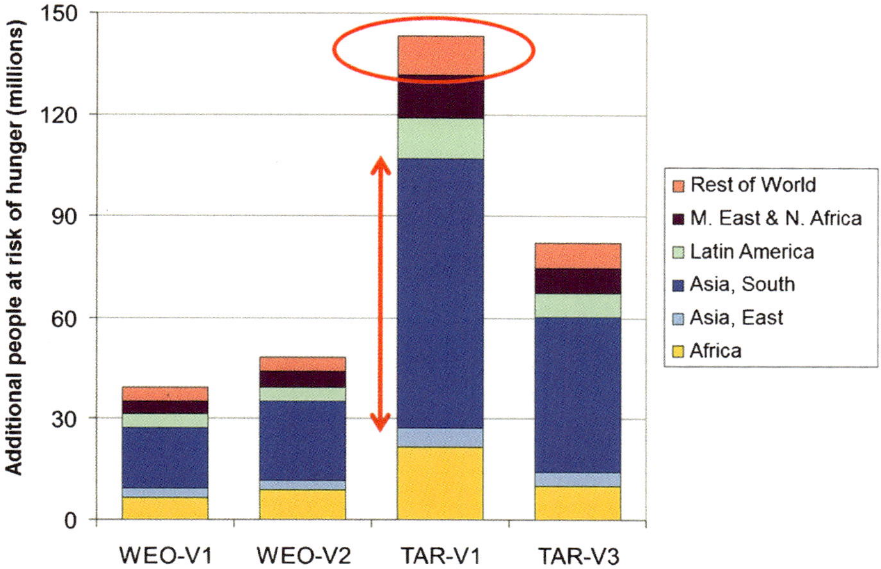

Fig. 10 Additional number of people at risk of hunger in 2020 relative to the reference scenario (Source: OFID and IIASA 2009)

This situation will inevitably lead to widespread hunger. By 2020 a further 140 million people will suffer from hunger and poverty because of the use of cereals to produce biofuels (see Fig. 10), most of them in South Asia (mainly India) and of course in Africa. The measures being developed to mitigate the risks of biofuel farming will undoubtedly reduce the threat, but in practice it will be impossible to eliminate the effects of biofuels on the supply and prices of food. There has been a long-running debate about feeding cereals to livestock when one-sixth of the world's population continues to suffer from chronic hunger and debilitating poverty. Now we are in danger of creating a situation where we end up for the next 30 years debating the immorality of feeding cereals to refineries.

Cereals are a very important part of the diets of millions of people in South Asia and Africa, and they will be most affected by the increased production of biofuel crops, even after second-generation biofuels are introduced. Quite the opposite will be true in developed countries due to the differences in wealth and diets, which depend far less on cereals. As a result, the levels of hunger will be lower there, despite the higher cereal prices. This is tantamount to forcing poorer countries to shoulder a large proportion of the cost of energy that more developed countries will enjoy.

Another factor that will have negative effects on the environment is the increased rate of deforestation (see Fig. 11). According to the IIASA model, an additional 18 million hectares of forest will need to be cleared for biofuel cultivation, mainly in Africa and Latin America. The latter is of course home to the Amazon rainforest, which is an integral part of the global ecosystem and is already under threat as large

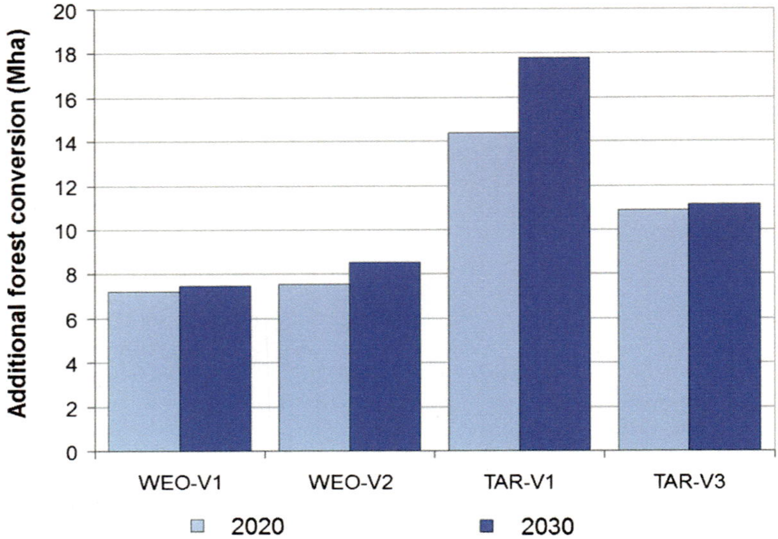

Fig. 11 Additional areas of forest converted to biofuel production relative to the reference scenario (Source: OFID and IIASA 2009)

Table 2 The impacts of biofuels: summary of the main results

Upward pressure on world food prices:	+30–50%
A factor in rising hunger:	+140 million people
Absorbing cereal production:	+260 million tonnes
Benefits for rural development:	+3–8% agricultural value added per annum
Mitigating climate change:	12.4 Gt CO_2 equivalent in 50 years
Competition with arable land:	−30–45 million ha
Fuelling deforestation:	+15–18 million ha
The fertiliser dilemma:	+10 million tonnes
A threat to biodiversity	Risks and opportunities
Energy security	6–12% biofuels in transport fuel

areas are already being cleared for crop and livestock production. Biofuels will only increase the demand for the rich land available there and contribute to deforestation.

4 Conclusions

The main findings of the study, summarised in Table 2, demonstrate that in setting biofuel targets, governments do not consider the full range of implications, including social and environmental externalities (see Hirschberg 2012). One example is the increase in hunger caused by a rise in world food prices of between 30% and 50%, and the use of about 260 million tonnes of cereals to produce biofuels.

The benefits of biofuels are minimal, amounting to an increase of around 3–8% in agricultural value added in developed countries. The expected benefit in terms of reduced GHG emissions and mitigating climate change is estimated to amount to 12.4 Gt of CO_2 equivalent by 2050. These benefits will come at significant cost, in particular the loss of 30–45 million hectares of arable land that will be converted to biofuel production, and the additional 10 million tonnes of fertiliser that will be needed, which will in turn pollute rivers and lakes.

To put the findings of the OFID and IIASA (2009) study into perspective, it should be remembered that biofuel targets aim to replace only 6–12% of the fuel used in the transportation sector. This then raises the critical question of whether these targets are worth it. As discussed in this chapter, considering all the variables involved, it is difficult not to be at least very cautious about the supposed benefits associated with biofuels in view of the sizeable costs involved. Were there to be more substantial reductions in CO_2 emissions, then the argument might be less skewed. The resources, especially water and land, that would need to be diverted to the cultivation and refining of biofuel crops in order to achieve a meaningful level of production, suggest that the returns, both financial and environmental, will be limited.

References

EC (European Commission). (2008). *Proposal for a directive of the European Parliament and of the Council on the promotion of the use of energy from renewable sources* (COM (2008) 19 final). Brussels: EC.

Hirschberg, S. (2012). Externalities in the global energy system. In F. L. Toth (Ed.), *Energy for development: Resources, technologies, environment* (pp. 121–138). Dordrecht: Springer.

IEA (International Energy Agency). (2008). *World energy outlook 2008*. Paris: OECD.

IPCC (Intergovernmental Panel on Climate Change). (2006). *IPCC guidelines for national greenhouse gas inventories*. Prepared by the National Greenhouse Gas Inventories Programme, Institute for Global Environmental Strategies (IGES), Hayama, Kanagawa, Japan. www.ipcc-nggip.iges.or.jp/public/2006gl/index.html

Licht, F. O. (2007). *World ethanol & biofuels report*. London: Agra Europe Ltd.

Licht, F. O. (2008). *World ethanol & biofuels report*. London: Agra Europe Ltd.

OFID, & IIASA (OPEC Fund for International Development, & International Institute for Applied Systems Analysis). (2009). *Biofuels and food security: Implications of accelerated biofuels production*. Vienna: OFID. www.ofid.org/publications/PDF/pamphlet/ofid_pam38_Biofuels.pdf

World Bank. (2008, April 12). *Development Committee Press Briefing*, Spring Meetings 2008. Washington, DC. Available at: http://web.worldbank.org/WBSITE/EXTERNAL/NEWS/0,,contentMDK:21728519~pagePK:34370~piPK:34424~theSitePK:4607,00.html

Index

F.L. Toth (ed.), *Energy for Development: Resources, Technologies, Environment*, Environment & Policy 54, DOI 10.1007/978-94-007-4162-1,

Printed by Printforce, the Netherlands